MULTIPLE-BASE NUMBER SYSTEM

Theory and Applications

Circuits and Electrical Engineering Series

Series Editor
Wai-Kai Chen

MicroCMOS Design
Bang-Sup Song

Multiple-Base Number System: Theory and Applications
Vassil Dimitrov, Graham Jullien, and Roberto Muscedere

MULTIPLE-BASE NUMBER SYSTEM

Theory and Applications

Vassil Dimitrov • Graham Jullien
Roberto Muscedere

CRC Press
Taylor & Francis Group
Boca Raton London New York

CRC Press is an imprint of the
Taylor & Francis Group, an **informa** business

CRC Press
Taylor & Francis Group
6000 Broken Sound Parkway NW, Suite 300
Boca Raton, FL 33487-2742

First issued in paperback 2017

Version Date: 20111209

ISBN 13: 978-1-138-07651-8 (pbk)
ISBN 13: 978-1-4398-3046-8 (hbk)

Visit the Taylor & Francis Web site at
http://www.taylorandfrancis.com

and the CRC Press Web site at
http://www.crcpress.com

Contents

Preface

This is a book about a new number representation that has interesting properties for special applications. It is appropriately catalogued in the area of computer arithmetic, which, as the name suggests, is about arithmetic that is appropriate for implementing on calculating machines. These machines have changed over the millennia that humans have been building aids to performing arithmetic calculations. At the present time, arithmetic processors are buried in the architectural structures of computer processors, built mostly out of silicon, with a minimum lateral component spacing of the order of a few tens of nanometers, and vertical spacing down to just a few atoms.

Arithmetic is one of the fields that even young children learn about. Counting with the natural numbers (1, 2, 3 …) leads to learning to add and multiply. Negative numbers and the concept of zero lead to expanding the natural numbers to the integers (…, –3, –2, –1, 0, 1, 2, 3 …), and learning about division leads to fractions and the rational numbers. When we perform arithmetic "longhand" we use a positional number representation with a radix of 10—undoubtedly developed from the fact that humans have a total of 10 digits on their two hands. Early mechanical as well as some electronic digital computers maintained the radix of 10, but the two-state nature of digital logic gates and storage technology leads to a radix of 2 as being more natural for electronic machines. Binary number representations, which use a fixed radix of 2, are ubiquitous in the field of computer arithmetic, and there are many valuable textbooks that cover the special arithmetic hardware circuits and processing blocks that make use of binary representations.

The subject of computer arithmetic is fascinating since it deals with the basic computational underpinnings of processor functions in all digital electronic systems with applications that are ubiquitous in our modern technological world. In fact, computer arithmetic is so fundamental to digital design that there is a danger of it being dismissed by many designers, who see it as well-worn knowledge that is already encased in hardware (actual or in a design library) or that can be readily called up from a variety of software packages. What is often missing, however, is that computer arithmetic can be optimized based on targeted applications and technologies, and easily available standard arithmetic hardware may not be the most efficient implementation strategy. There is also a strong curiosity element associated with unusual approaches to implementing fundamental computer operations, and in this book we try to stir in a mixture of curiosity with our pragmatic approaches to implementing basic arithmetic for two application areas: digital signal processing and cryptography. As such, this book is somewhat different from most of the textbooks on computer arithmetic, in that it deals almost entirely with a rather new multiple-base number system that is rapidly gaining in popularity. In its original form, the double-base number system (DBNS) can be viewed as an extension to the binary number system. By using a second base, in addition to the base 2 used by the binary number system, we uncover some rather remarkable properties of both the DBNS and its logarithmic extension, the multidimensional logarithmic number system (MDLNS).

We trust that the reader will find something fascinating about multiple bases and, if we have done our job well, will also be convinced that there are applications that can benefit from, at least, a serious consideration of this number representation and the techniques we have identified to efficiently implement calculations using multiple bases. The authors of the book have a collective interest in pursuing alternative representations that have the potential to improve aspects of the implementation of high-performance calculations in hardware. Two of us are engineers (Jullien and Muscedere), and the principal author, Dimitrov, has both a mathematics and an engineering background. All three of us are dedicated to implementing our discoveries in advanced integrated circuit technologies, and some of these designs are used to illustrate the theory and architectural techniques that are disclosed in this book.

In summary, the purpose of the book is to showcase the usefulness of the multiple-base number representation in various applications, and our main goal is to disseminate the results of our research work to as wide as possible an audience of engineers, computer scientists, and mathematicians. We hope this book at least partially achieves this goal.

Vassil S. Dimitrov
Graham Jullien
Roberto Muscedere

About the Authors

Vassil S. Dimitrov earned a PhD degree in applied mathematics from the Bulgarian Academy of Sciences in 1995. Since 1995, he has held postdoctoral positions at the University of Windsor, Ontario (1996–1998) and Helsinki University of Technology (1999–2000). From 1998 to 1999, he worked as a research scientist for Cigital, Dulles, Virginia (formerly known as Reliable Software Technology), where he conducted research on different cryptanalysis problems. Since 2001, he has been an associate professor in the Department of Electrical and Computer Engineering, Schulich School of Engineering, University of Calgary, Alberta. His main research areas include implementation of cryptographic protocols, number theoretic algorithms, computational complexity, image processing and compression, and related topics.

Graham Jullien recently retired as the iCORE chair in Advanced Technology Information Processing Systems, and the director of the ATIPS Laboratories, in the Department of Electrical and Computer Engineering, Schulich School of Engineering, at the University of Calgary. His long-term research interests are in the areas of integrated circuits (including SoC), VLSI signal processing, computer arithmetic, high-performance parallel architectures, and number theoretic techniques. Since taking up his chair position at Calgary, he expanded his research interests to include security systems, nanoelectronic technologies, and biomedical systems. He was also instrumental, along with his colleagues, in developing an integration laboratory cluster to explore next-generation integrated microsystems. Dr. Jullien is a fellow of the Royal Society of Canada, a life fellow of the IEEE, a fellow of the Engineering Institute of Canada, and until recently, was a member of the boards of directors of DALSA Corp., CMC Microsystems, and Micronet R&D. He has published more than 400 papers in refereed technical journals and conference proceedings, and has served on the organizing and program committees of many international conferences and workshops over the past 35 years. Most recently he was the general chair for the IEEE International Symposium on Computer Arithmetic in Montpellier in 2007, and was guest coeditor of the IEEE Proceedings special issue *System-on-Chip: Integration and Packaging*, June 2006.

Roberto Muscedere received his BASc degree in 1996, MASc degree in 1999, and PhD in 2003, all from the University of Windsor in electrical engineering. During this time he also managed the microelectronics computing environment at the Research Centre for Integrated Microsystems (formally the VLSI Research Group) at the University of Windsor. He is currently an associate professor in the Electrical and Computer Engineering Department at the University of Windsor. His research areas include the implementation of high-performance and low-power VLSI circuits, full and semicustom VLSI design, computer arithmetic, HDL synthesis, digital signal processing, and embedded systems. Dr. Muscedere has been a member of the IEEE since 1994.

1

Technology, Applications, and Computation

1.1 Introduction

The field of computer arithmetic is a fascinating subject, and not at all the drudgery that most of us experienced with our first exposure to the third *r* of the three *r*'s at school. The first two *r*'s are related as requirements of the processes required to become literate. The third *r* relates to numeracy, the understanding and implementation of computation which is a basic requirement for "technical literacy," as important a skill in today's world as was traditional literacy in the past. In this introductory chapter, we provide a brief history and introduction to number systems and machine calculation; we emphasize special applications that drive the search for efficient machine arithmetic given the requirements of the applications and the available technology.

1.2 Ancient Roots

Ancient numbering systems abound, but the most striking is the system developed by the Babylonians starting around 5,000 years ago. It is striking in that it is a form of weighted positional system, which we use today. However, the Babylonians did not have a symbol for zero (instead they used a space) and the weighting was based on the number 60 rather than the number 10, which we use today for representing decimal numbers. (Vestiges of the weight 60 can still be found in the way we measure time and angles.) It also appears that the binary number system, which we naturally think of as being introduced with the advent of electronic computers built with logic gates, was used at least 4,000 years ago for determining weights using a simple balance and carefully weighed stones [1].

1.2.1 An Ancient Binary A/D Converter

We can imagine a trader from 4,000 years ago by the side of a river, setting up a balance and then searching for a handful of river-washed pebbles that had specific weight characteristics. These characteristics were determined from the simple operation of producing a balance between sets of stones. The technology used here was based only on the force of gravity using a balance bar and fulcrum. The operation is demonstrated in Figure 1.1 for an equivalent 4-bit representation (1–15 times the smallest stone weight). The accuracy of the number system is determined by the accuracy with which the stones were selected and correctly balanced (including positioning the stones so that their accumulated center of gravity was always in the same position on the opposite sides of the balance). The relative weights of the stones in the full measurement set are shown in Figure 1.1 as {1, 1, 2, 4, 8}. Designers of binary-weighted digital/analog converters (D/As) know this sequence well! Such a D/A converter can be used to implement a successive approximations A/D converter. The two 1's are redundant in terms of a 4-bit measurement system, only being required to generate the full measurement set. In a sense, the traders of 4,000 years ago had also built an A/D converter in which an analog commodity weight was converted into a subset of stones from the full measurement set.

1.2.2 Ancient Computational Aids

Computational aids and calculators also have ancient roots. Counting boards (a precursor of the "more modern" abacus using beads) have been used for several thousand years [2], with some evidence that they were

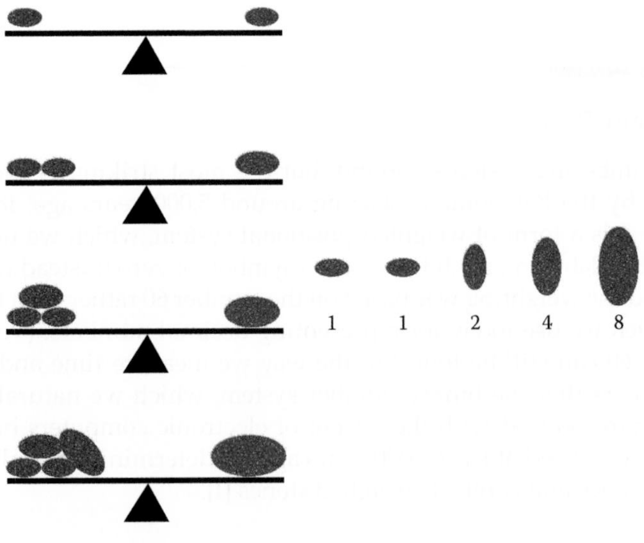

FIGURE 1.1 (See color insert)
Four-thousand-year-old binary number system.

initially developed by the Romans. Of some surprise, an astronomical prediction machine, circa 200 BC, was recovered from an ancient shipwreck off the Greek island of Antikythera in 1901. This mechanism was truly advanced based on the fact that mechanical calculators of similar complexity did not appear again (or, at least, have not been found) for at least another 1,500 years. Based on an analysis of CT scans of the components [3], the calculator used a variety of sophisticated precision gears, each with up to several hundred teeth, along with epicyclic gearing. The mechanism was able to compute addition, subtraction, and multiplication with precise fractions and, for example, could predict the position of the moon based on models available at the time.

1.3 Analog or Digital?

Over the past five centuries the interest in computational aids and calculating machines has steadily increased, with an explosive growth over the past 100 years, driven by mechanical devices, electromechanical devices, and electronics.

Mechanical and electromechanical devices were based on decimal arithmetic because of the need to interact with human operators. Some of the early electronic computers also used the decimal-based number systems for the same reason, even though the two-state property (0, 1) of signal propagation into and out of computer logic circuits provides a perfect match with pure binary number representations. Until the latter part of the 20th century, analog devices were also heavily used for advanced mathematical computations, such as finding solutions of nonlinear differential equations, where a physical quantity (e.g., voltage in electronic circuits or rotation in mechanical systems) is observed as the solution to a problem. In this case there is no implied number system, only that of the operator in interpreting the analog results. Analog systems have a relatively large computational error (several percent) compared to the much higher precision capable of digital machines, but there were application areas found for both analog and digital machines. We will discuss an analog method of computing with a double-base system in Chapter 3.

Examples of applications that took advantage of the disparity in complexity and error between analog and digital mechanisms are ideally portrayed in comparing Lord Kelvin's analog tide prediction machine [4] to Babbage's digital difference engine [5]; they were proposed within a few decades of each other in the 19th century.

1.3.1 An Analog Computer Fourier Analysis Tide Predictor

The tide predictor computed the height of the tide from a datum point as a function of time using the form of Equation (1.1):

$$h(t) = H_0 + \sum_{i=1}^{10} \{A_i \cos[g_i t + f_i]\} \tag{1.1}$$

The predictor combines tidal harmonic constituents, based on harmonic components of the changes of the position of the earth, sun, and moon, to perform the prediction. The technique is still used today to make tide predictions, though the 10 components in Lord Kelvin's first machine are augmented by a factor of 4 or more, based on location [4]. This analog computer used a system of pulleys, wires, and dials (see Figure 1.2), and the generation of sinusoid motion used a rotating wheel with an

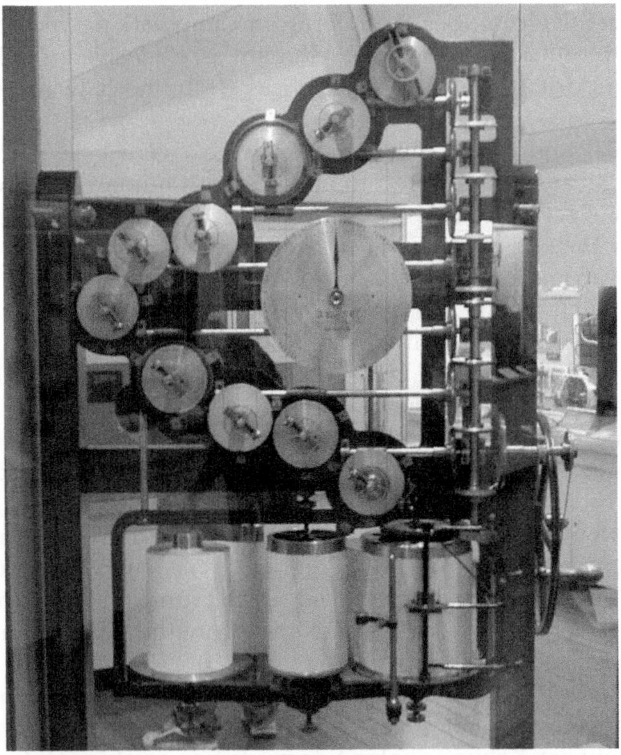

FIGURE 1.2 (See color insert)
Lord Kelvin's tide predicting machine (1876), Science Museum, London. (Copyright 2005 William M. Connolley under the GNU Free Documentation License.)

off-center peg to drive a vertically moving arm (the inverse of a piston driving a crankshaft), and similar mechanisms can also be seen in the Antikythera machine of 2,000 years earlier! Lord Kelvin's predictor could print out calculations of harbor tide patterns for up to a year in about 4 hours by turning a crank. Clearly errors of a few percent were acceptable for this application, particularly considering the errors in the astronomical models used.

1.3.2 Babbage's Difference Engine

Charles Babbage's difference engine [5] was designed to tabulate the values of polynomials of *n*th degree using the method of finite differences. The engine design used mechanical wheels and gears to compute using the decimal number system; the main purpose of the engine was to produce accurate mathematical tables with no human error (including direct printing from the engine). His difference engine no. 2 was able to hold (store) eight numbers of 31 decimal digits each, allowing the machine to tabulate seventh degree polynomials to that precision. By using Newton's method of divided differences, the engine can compute a sequence of values of a polynomial with only the need to add and store, using a *ten's complement* form for negative numbers. There were some intriguing aspects to the design of the engine, including the addition and carry propagation, using carry save and carry skip techniques, techniques that are still used in modern microcomputers. Babbage never completed the building of the difference engine; we had to wait until the 1990s to be able to see replicas at the Science Museum in London and the Computer History Museum in Mountain View, California (see Figure 1.3).

1.3.3 Pros and Cons

Comparing the tide predictor to the difference engine, we note that the difference engine would not be able to automatically compute the complete tide prediction model of Equation (1.1); however, the harmonic components could be approximated by finite order polynomials so parts of the equation could be individually evaluated, with the final summation to be carried out at the end. Perhaps the analytical engine [5,6] that Babbage proposed after abandoning the construction of the difference engine would have been more suitable based on its programmability with punched cards. However, the complete analytical engine has never been built, and even if it were it would be so cumbersome as to require a real engine to turn the crank! Analog computing machines were clearly much easier to build than digital machines using the technology of the 19th century, and this was still the case up to the end of the vacuum tube era (middle of the 20th century).

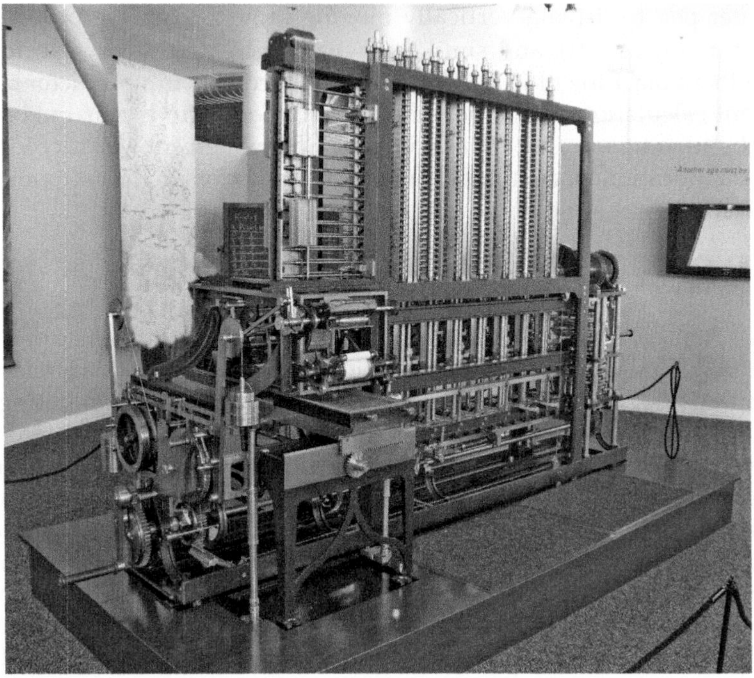

FIGURE 1.3 (See color insert)
Babbage's difference engine, Computer History Museum, Mountain View, California. (Copyright 2009 Allan J. Cronin under the GNU Free Documentation License.)

1.3.4 The Slide Rule: An Analog Logarithmic Processor

Analog and digital computational aids were available to students and engineers long before the pocket calculator arrived. One of the authors went through his entire college and university education using tables of logarithms and trigonometric functions (digital aids) and a slide rule (analog aid). The $6.50 purchase of the 1,046-page standards book edited by Abramowitz and Stegun [7] was a particularly good deal for the professional engineer in 1969! The tables of logarithms (and their inverses) were mainly used for accurate computations of multiplications (including reciprocals and powers), with the slide rule being used where lower accuracy could be tolerated—this meant most of the time! The slide rule uses logarithmic scales on both the static body and the slide, so clearly, in both the digital and analog approaches, we are using the mapping property of multiplication to addition in going from a linear scale to a log scale. For the tables we looked up the logarithms of the multiplier and multiplicand, added the logarithms, and looked up the inverse logarithm in the tables. It was assumed, of course, that we were skilled at adding numbers. For the slide rule calculations the logarithmic map was automatically performed with the logarithmic scale. The addition was performed by concatenating the slide and body markings and looking up the sum, as shown in Figure 1.4 (using the author's slide rule—warts and all). The figure shows the cursor "looking up" the concatenation of the body scale at 1.5 with the slide scale at 2.5, giving a result of between 3.74 and 3.76 (the accurate answer is, of course, 3.75).

The master of the slide rule knew the trick to compute $\sqrt{a^2+b^2}$ in three moves without having to do a difficult addition. The trick was to rewrite the expression as shown in Equation (1.2):

$$\sqrt{a^2+b^2} = a \cdot \sqrt{1+\left(\frac{b}{a}\right)^2} \tag{1.2}$$

Computing the right-hand side of Equation (1.2) involves a division, a squaring, adding 1, a square root, and a multiplication. Assuming the slide rule has a scale of squares (all good slide rules did), then the only tricky move was the addition of 1, which is trivial.

Figure 1.5 shows the three moves used to compute $\sqrt{3^2+4^2} = 5$. We elect to choose $a > b$ so that $b^2/a^2 < 1$ and the addition will yield a number with 1 to the left of the decimal point. From Figure 1.5(a) we see the first move, where we divide 3 by 4 (0.75) by subtracting the scales and then looking up the square (~0.562) on the right-hand side of the scale (at 100 on the x^2 scale). We now move the slide to 1.562 (or as close as possible) on the left-hand side of the slide and move the cursor to 4 on the x scale of the slide. Looking at the x scale on the body of the slide rule, we get the answer, 5. The trick here was to rewrite the expression so that the addition is trivial. This same idea is used in logarithmic arithmetic to turn addition with two addends into a unary lookup table. In our group we affectionately refer to this as the *slide rule trick,* and will so refer to it in subsequent chapters dealing with a multiple-base logarithmic arithmetic. As an aside, we note that we have to keep track of the magnitude of the numbers in the calculation. Clearly the 56.2 value in Figure 1.5(a) has to be interpreted as 0.562 so that the slide is located at 1.562 in Figure 1.5(b). Keeping track of magnitudes of calculations

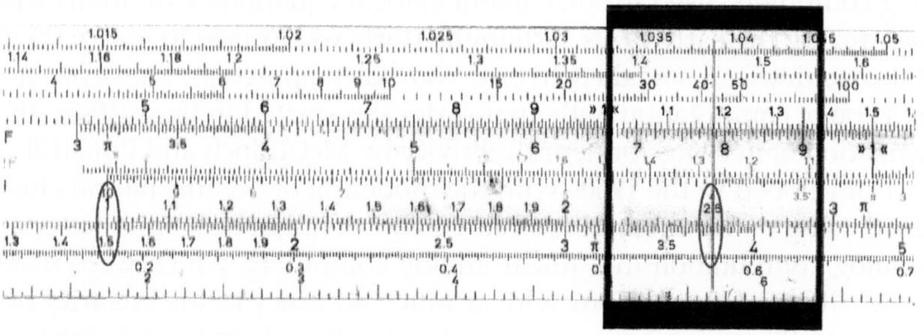

FIGURE 1.4 (See color insert)
Multiplication of 1.5 by 2.5 on a slide rule.

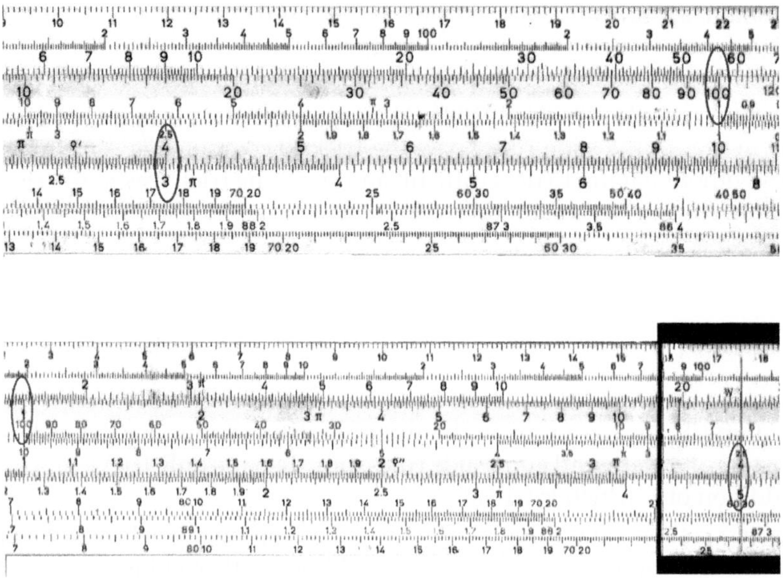

FIGURE 1.5 (See color insert)
Three moves to compute the slide rule trick.

is a skill that has seen some demise since the availability of the ubiquitous pocket calculator!

It is interesting to note that logarithmic arithmetic was suggested as a computational implementation tool for digital filters as far back as 1971 by Kingsbury and Rayner [8], with considerable interest since then in applying logarithmic arithmetic to this application area [9–11].

1.4 Where Are We Now?

Analog computers, used to solve mathematical equations, essentially disappeared during the early 1970s, although efforts were made to marry them to digital machines—a so-called hybrid computer approach. However, a different type of computation with analog devices started in the 1940s with the investigation of nervous (or neural) activity by McCulloch and Pitts [12], and further developed in the 1960s with the modeling of organic neural clusters (brains!) using nonlinear analog circuits. Thus it might be helpful to separate out analog computation into linear analog computers, which were used to evaluate differential equations with as much accuracy as the linearity of the amplifiers with passive component feedback allowed, and nonlinear analog processors, which include neural networks, cellular nonlinear networks, and their analog circuit implementations.

1.4.1 Moore's Law

The 1970s also saw the birth of the microprocessor and the start of the current revolution in high-performance computational systems. The advent of integrated transistor circuits (ICs) has allowed for an exponential increase in the complexity of single-chip computational circuits and the attendant advances in information processing capability. Moore's law [13] has stood the test of time (4½ decades and counting) in spite of the need for regularly finding solutions to major roadblocks in lithography and other aspects of IC fabrication technology. The predictions of an imminent departure from Moore's law abound throughout the decades of advances in IC technology. Though as we approach atomic dimensions in the active devices built on silicon, Moore's law certainly seems threatened, unless we can find new nanotechnologies to provide a continuation of the law. Advances in IC fabrication density and device speed have had a profound effect on the basic computational units that are present in almost every IC that is fabricated today. In fact, there has been a revolution in the number and type of new applications that have appeared simply because of the availability of billions of active devices on a single sliver of monolithic crystalline silicon.

The traditional applications for computing, such as the physical sciences, finance, and military applications, drove the first forays into electronic computers and have benefited enormously from the improvements, through vacuum tubes, transistors, and integrated circuits. For these applications, the requirements for computational accuracy have driven the design of the arithmetic units, in particular the use of floating point arithmetic as a way of representing numbers with precision and dynamic range, in spite of a relatively large overhead in the hardware required to perform the arithmetic. The demand for faster and more accurate processing power also led to the birth of supercomputers in the 1960s. These are computers that are able to process orders of magnitude more information per second than more accessible machines. A definition of number-crunching processing power, *floating point operations per second* (FLOPS), was also introduced, and we have seen this measure increase from mega-FLOPS to peta-FLOPS over just a few decades. Interestingly, even cheap consumer computers exceed the definition of a supercomputer of just a few years earlier, often faster than the U.S. military changes the definition! The ubiquitous FLOP has also been used as a more general measure of computer processing power, even if the main use is not continuous computation with floating point arithmetic. In fact, all of a computer's systems, including its architecture, in addition to arithmetic units, need to have performance improvements in order to provide sustained data to the arithmetic unit(s).

1.4.2 The Microprocessor and Microcontroller

The first microprocessors appeared in 1970–1971: the most well known is the 4004 from Intel Corp. [14]. The CPU chip was actually part of a four-chip set used to build a complete computer system. By the mid-1970s, complete computer systems on a single chip were available in which all of the components required to build a complete computer system (albeit very limited) were contained on the chip. One of the first chips was the Intel 8748, which contained an erasable programmable read-only memory (EPROM) for programming, and most of the pins were available as input/output (I/O) lines. Along with programmable I/O pins, the 8748 contained a basic timer and interrupt structure that together would normally be considered nonessential for a standard computer system, let alone for a resource-limited single chip of that time. In fact, this and subsequent complete single-chip computers were targeted to a new market where the computer was embedded into a product that was not, itself, a computer. In reality, the chip was a very flexible microcontroller. As an example, one of the major applications for the 8748 was as a keyboard controller for IBM computers. The terms *microcontroller* and *embedded systems* are very familiar to us now, but their birth was with the development of devices such as the 8748. In terms of being a computational unit, they were very limited by today's standards, as were all of the microprocessor products of that time, but that did not stop users of these devices from exploring ways that fast computations could be carried out using alternative techniques [15].

The first desktop computers, based on microprocessors, started to appear in the late 1970s. They coexisted with mini-computers for several years until they became sufficiently powerful so that they could take on the same complexity of computational tasks that mini-computers had handled just a few years previously. In fact, whatever had been on a 19-inch rack-mount-printed circuit board in the mini-computer era was now available on one or two very large-scale integration (VSLI) chips. These included coprocessors that were expressly designed to offload floating point and integer arithmetic calculations from the main processor.

1.4.3 Advances in Integrated Circuit Technology

Advances in the implementation of computation have been driven by advances in integrated circuit technology and the evolution of the approaches to building computational circuits. The technology advances were predicted by Gordon Moore [13] with remarkable accuracy, though the fact that the industry uses Moore's law as a guide to its own technology improvements [16] perhaps introduces something of a self-fulfilling prophecy into the picture. The semiconductor industry is very inventive, and there have been several major breakthroughs that have maintained the exponential advances, often against the predictions of top people in the field. The first microprocessors were built with p-channel metal oxide semiconductor (PMOS) and then n-channel MOS (NMOS) technologies, but the star technology is still complementary metal oxide semiconductor (CMOS). CMOS became the dominant MOS digital technology in the 1980s, and has maintained this position in spite of other technologies, such as Bipolar-CMOS (BiCMOS) and GaAs, that, for a time, offered some advantages, e.g., speed, over CMOS. In reality, CMOS is a set of logic families that make use of the ability to produce complementary devices in the same process, and we have seen a wide variety of static and dynamic CMOS logic families that have been used to implement computation. The first commercial CMOS processes used lithography (the optical technique of patterning photosensitive material on semiconductor substrates) with dimensions measured in microns. At the time of writing this book, 22 nm processes are starting to come on line, and 45 nm ICs are in the marketplace. An experimental (in 2009) Intel microprocessor family has 48 cores (individual processors) on a 1.3 billion transistor single chip (called the cloud computer). The computational power of such chips is measured in tera-FLOPS.

1.5 Arithmetic and DSP

Concurrent with the commercial development of integrated circuits in the 1960s, a new field of research was beginning to emerge: digital signal processing (DSP). In DSP, signals are sampled both in time (discrete time) and in magnitude (digital). Although the basic theoretical techniques were quite well known before the commercial development took place, the use of DSP only took a firm foothold with the advent of semiconductor electronics and, in particular, IC fabrication. We will spend a few pages discussing this subject because the search for efficient computational techniques for DSP algorithms led to an increased interest in alternative arithmetic techniques, including the double-base number system. In this section, for completeness, we review the mathematical bases for replacing classical analog processing with DSP.

1.5.1 Data Stream Processing

One of the features of many DSP algorithms is the requirement for data stream processing, that is, the continuous application of a computational algorithm on an input digital data stream to produce a continuous output data stream. This is exactly what analog signal processing does, but analog computers operate on continuous waveforms, and instead of digital logic, analog processors use amplifiers, resistors, capacitors, and inductors. Replacing continuous (analog) signal processing with DSP has some limitations, but these are overcome in practice. This represents a fundamental difference between the use of computers to solve a problem, where a program is written to input numerical data and to output the solution to the problem as another set of data. Once the solution has been obtained, the computer has done its job! In data streaming, however, the processor is continuously fed with a stream of numbers (e.g., a regularly sampled analog signal), and outputs a stream of numbers at the same rate as the input. The stream of input numbers may be essentially infinite (e.g., samples of the output of a microphone that is never turned off), so the processor is never finished. The throughput rate (the inverse of the time between adjacent samples of the waveform) is determined by the application, and early DSP hardware was woefully inadequate to keep up with even the lowest-bandwidth applications. During the final three decades of the 20th century, many different processing techniques, including alternative arithmetic techniques and special architectures, were developed to squeeze higher and higher throughput rates out of the available hardware. As IC technology advances, mainstream computer arithmetic solutions can operate at sufficiently high-throughput rates to satisfy many of the ubiquitous applications, such as multimedia streaming, with acceptable results. However, as the definition of *acceptable* becomes more and more severe, as new applications take hold, and as we approach a potential brick wall for current technologies, it will be very useful to maintain interest in alternative techniques in order to increase the degrees of freedom to the designers of future systems.

1.5.2 Sampling and Conversion

The sampling operation allows continuous waveforms to be captured at regular intervals as streams of digital data. Figure 1.6 demonstrates the effect of sampling continuous signals by looking at the time and frequency domain of a sampled band-limited time domain waveform using the Fourier transform (FT) [17]. The figure shows time domain waveforms on the left and frequency domain waveforms on the right.

In Figure 1.6 we make use of the *multiplication* ⇔ *convolution* FT mapping property. The FT is shown in Equation (1.3), where $f(t)$ is an integrable function, t is the variable in the time domain, and ω is the variable in the radian frequency domain; i is the complex operator.

$$F(\omega) = \int_{-\infty}^{\infty} f(t) e^{-i\omega t} dt \tag{1.3}$$

Equation (1.4) shows the inverse FT.

$$f(t) = \int_{-\infty}^{\infty} F(\omega) e^{i\omega t} d\omega \tag{1.4}$$

The FT is used to analyze the frequency domain components of a time domain waveform. The multiplication ⇔ convolution mapping property operates in both the forward and inverse transform directions; Figure 1.6 uses the forward mapping direction. We also use the property that a periodic

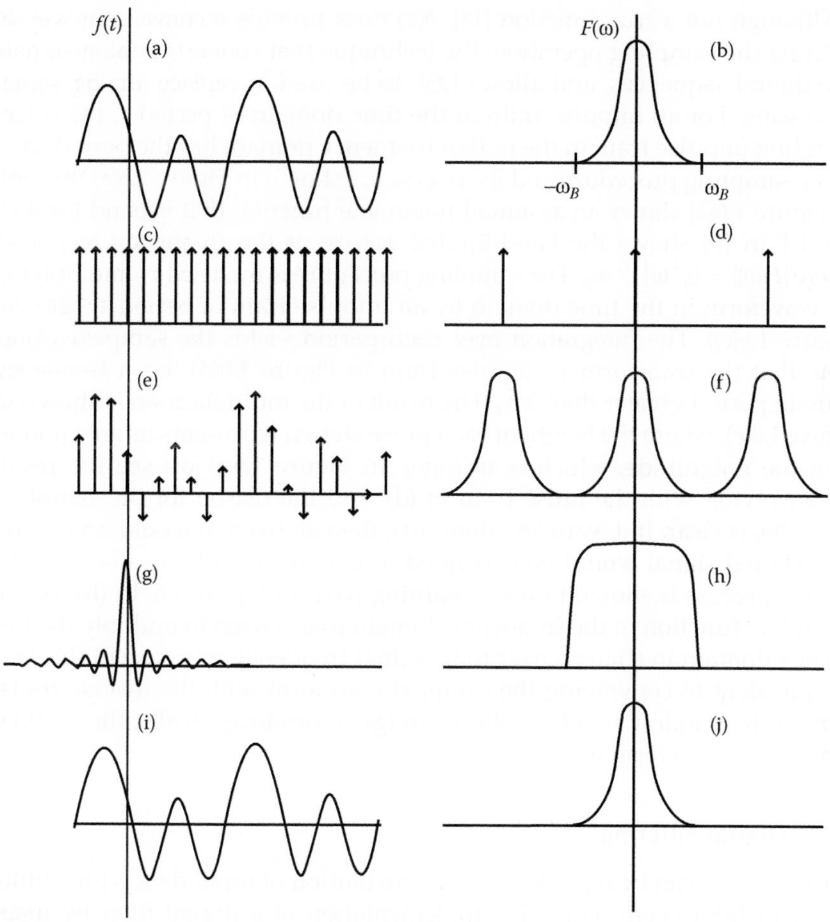

FIGURE 1.6
Processing continuous signals using digital signal processing (DSP).

train of Dirac delta functions (impulse train) in the time domain maps to another impulse train in the frequency domain [17]. A Dirac delta function, $\delta(x)$, is defined as in Equation (1.5):

$$\delta(x) = \begin{cases} +\infty, & x = 0 \\ 0, & x \neq 0 \end{cases} \tag{1.5}$$

where $\int_{-\infty}^{\infty} \delta(x)\,dx = 1$.

Although not a true function [18], $\delta(x)$ does provide a convenient way of defining the sampling operation: the technique that converts analog signals into digital sequences, and allows DSP to be used to replace analog signal processing. For an impulse train in the time domain of period τ, the corresponding impulse train in the radian frequency domain has the period $2\pi/\tau$.

The sampling procedure and its inverse are shown in Figure 1.6(a) through (j). Figure 1.6(a) shows an assumed integrable function that is band limited. The FT in (b) shows the band-limited nature of the frequency response, where $|F(\omega)| = 0, |\omega| > \omega_B$. The sampling procedure is modeled by multiplying the waveform in the time domain by an impulse train of period $\tau \leq 2\pi/2\omega_B$ (Figure 1.6(c)). The integration over each period yields the sampled value. Note that the transformed impulse train in Figure 1.6(d) has a frequency domain period greater than $2\omega_B$. The result of the multiplication is shown in Figure 1.6(e), where the height of each pulse shown represents its area (rather than the magnitude, which is infinite). In Figure 1.6(f) we see the result of convolving with the pulse train in (d), and the reason for the sampling rate $> 2\omega_B$ is clear. If it were less than this, then an overlap would occur, and the original signal would be corrupted; this is referred to as aliasing. The inverse process is shown in the remaining parts of Figure 1.6. In (h) we see a window function in the frequency domain that is used to multiply the frequency domain in (f) to recover the original frequency response in (b). This is equivalent to convolving the sampled waveform with the inverse transform of the window function, shown in (g), to produce, ideally, the original continuous waveform in (i).

1.5.3 Digital Filtering

Since filtering can be regarded as the convolution of input data with a finite vector of filter coefficients, the implementation of a digital filter by mapping the input signal to the DFT domain, multiplying by the previously mapped filter coefficients, and then inverting the result, could be performed very efficiently using one of the many developed *fast Fourier transform* (FFT) algorithms [19]. Other early techniques for digital filtering used recursive structures in which the output of the filter is returned to the input through delay elements resulting in a theoretically infinite response to an impulse input function. These two techniques are often referred to as finite impulse response (FIR) and infinite impulse response (IIR) filters, respectively. In the early years following the publishing of the first FFT algorithms [20], FIR filtering using the FFT was mostly performed offline, because of the need to compute the filtering in blocks, and direct implementations of useful FIR filters were computationally intensive. On the other hand, recursive filters could be used with a considerable reduction in the number of arithmetic operations per output sample, leading to real-time implementations with existing semiconductor technology. Early custom hardware used integer arithmetic and IIR filters for real-time implementations, where custom scaling approaches were used to prevent overflow of the data registers. This involved a careful trade-off between arithmetic efficiency and problems due to nonlinearities associated with truncating intermediate arithmetic results, including limit cycle oscillations and overflow oscillations [21].

DSP computations are performed on the sampled waveform of Figure 1.6(e) before any conversion back to a continuous waveform is performed. A commonly used DSP algorithm is a digital filter, which involves many multiplications and additions operating on each input sample, and was an early target for innovative arithmetic techniques. Digital filters were among the first class of DSP algorithms to be researched and developed into commercial products. The early thrust of DSP can be found in the classic text that brought many of us into the field over 40 years ago [21]. The defining algorithm for early work on DSP is the discrete Fourier transform (DFT). The DFT has properties mirroring those of the Fourier transform, but in the sampled (discrete) time and frequency domains. An important property is that of circular convolution, which corresponds to the continuous convolution property we have briefly looked at earlier, but operates in the digital domain. This was arguably the first indirect architecture used for digital filtering, using the circular convolution property of the DFT and the arithmetic efficiencies brought about by the fast Fourier transform. We briefly discuss this in the following two sections.

1.6 Discrete Fourier Transform (DFT)

We devote an entire section on the DFT because the discovery of a set of fast algorithms to compute the DFT was an early driver for the commercialization of DSP.

If we look at Figure 1.6(e) and (f), we see that the effect of sampling in the time domain produces a periodic function in the radian frequency domain. If we now sample in the frequency domain, we will generate a periodic pulse train in the time domain; this is sketched out in Figure 1.7. Since both the radian frequency domain and time domain functions are periodic, we simply need to perform cyclic shifts around a single period in order to capture the information in adjacent periods.

Based on capturing information about the periodic sequences in only a single period, the DFT and its inverse (IDFT) are defined as in Equations (1.6) and (1.7):

$$X_k = \sum_{n=0}^{N-1} x_n \cdot e^{-i\frac{2\pi kn}{N}} \tag{1.6}$$

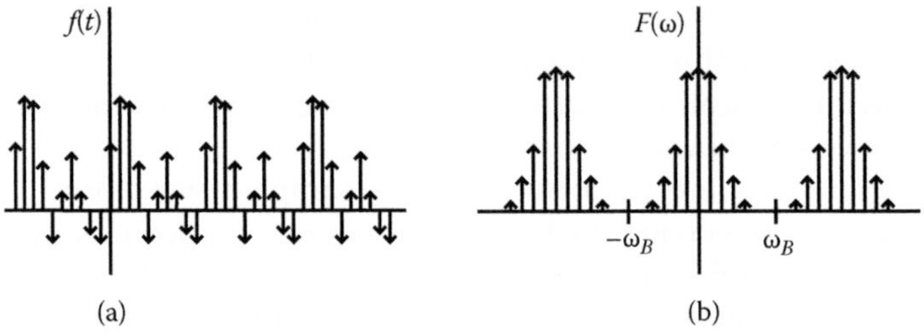

FIGURE 1.7
Sampling in the frequency domain.

$$x_n = \frac{1}{N}\sum_{k=0}^{N-1} X_k \cdot e^{i\frac{2\pi kn}{N}} \tag{1.7}$$

where x_n is the nth sample of the N sample input and X_k is the kth sample of the DFT output of N complex samples; i is the complex operator. The DFT can be regarded as a discrete approximation to the continuous Fourier transform (FT) of an integrable function; however, what is probably more important to the application of the DFT to digital filtering is that the transform has an exact inverse (we can substitute Equation (1.7) into Equation (1.6) to prove that), and that the properties of the FT are mirrored in the DFT, also as exact properties, but with a discrete sequence of N values and cyclic shifting. Thus, the convolution property of the FT is mirrored as a cyclic convolution property (where the shifted sample indices are computed over a finite ring with modulus N). The cyclic shifting property is a nuisance when what we really want is ordinary (acyclic) shifting, but tricks were developed to overcome this, for example, zero sample padding, that can map a subset of samples of cyclic convolution into linear convolution [21].

Why should we go to all of this trouble? Well, we need to look at the computational complexity associated with implementing convolution filters. The convolution of a data stream, x, with a filter impulse response, h, is given in Equation (1.8). Note the cyclic shift of the M coefficient impulse response "over" the input data stream, where $\langle m-n\rangle_M \equiv \langle m-n\rangle \bmod M$.

$$y_n = \sum_{m=0}^{M-1} x_m \cdot h_{\langle m-n\rangle_M} \tag{1.8}$$

The computational requirements for the direct computation of each sample are M multiplications and $M-1$ additions. Using a standard binary fixed-point representation with a standard shift-and-add multiplication algorithm, each multiplication requires $B-1$ additions, where B is the number of bits in the representation. It is clear that multiplication dominates the arithmetic cost (some combination of hardware and time delay), with a multiplicative complexity of $O(M^2)$ for computing M output samples. Indirectly computing the convolution sum of Equation (1.8) using the cyclic convolution property of the DFT appeared to be an exercise without merit; however, this view changed with the introduction of FFT algorithms. The commonly used FFT algorithms have a multiplicative complexity of $O(M \log_2 M)$. Clearly, for large values of M, even computing two forward transforms and one inverse transform will provide considerable reduction in the hardware (or time delay) cost of performing digital convolution. For completeness, a brief review of the FFT is contained in the following subsection.

1.6.1 Fast Fourier Transform (FFT)

DSP research and development undoubtedly received a large impetus with the publishing of an efficient algorithm [20] for computing the discrete Fourier transform (DFT), the so-called fast Fourier transform (FFT). Although many different algorithms were developed that are referred to as FFTs, the main algorithms still in use today are the original decimation in time (DIT) and decimation in frequency (DIF).

All of the FFT algorithms use the concept of reducing the multiplicative redundancy inherent in calculating the DFT directly. This redundancy is extracted and removed by forming subsequences of the summation in Equation (1.6) for the forward DFT, and the summation in Equation (1.7) for the inverse DFT [21]. The redundancy is eliminated when using a sample size that is a power of 2. For the case of a DIT reduction, the final computation of an 8-point DFT is presented as a flow graph in Figure 1.8.

The input to the nodes (filled circles) on the flow graph are summed at the node. The arrows on paths in the flow graph denote direction of the computation (which proceeds left to right) and variables beside the arrows represent data multipliers, where $W^n = e^{-i2n/N}$ (Nth roots of unity). Noting that $W^0 = 1$, we can, for example, trace back from the X_0 output to all of the inputs, and we indeed see that $X_0 = \sum_{i=0}^{7} x_i$, which is the correct result obtained from Equation (1.6). We will leave it up to the (dedicated) reader to trace back all of the other outputs! As a humorous aside, the W^n multipliers are affectionately referred to as *twiddle factors* [21]. We will look more closely at such flow graphs when we discuss applications and arithmetic. However, of note is the fact that the computations are with complex numbers, all of the replicated structures have two inputs and two outputs (radix 2), the only arithmetic operations are multiplication and addition (no division), and the algorithm is feed-forward only.

There was considerable research published in the 1970s and early 1980s on many different forms of fast DFT algorithms (all referred to as FFTs) with regard to improving arithmetic efficiency and flow graph

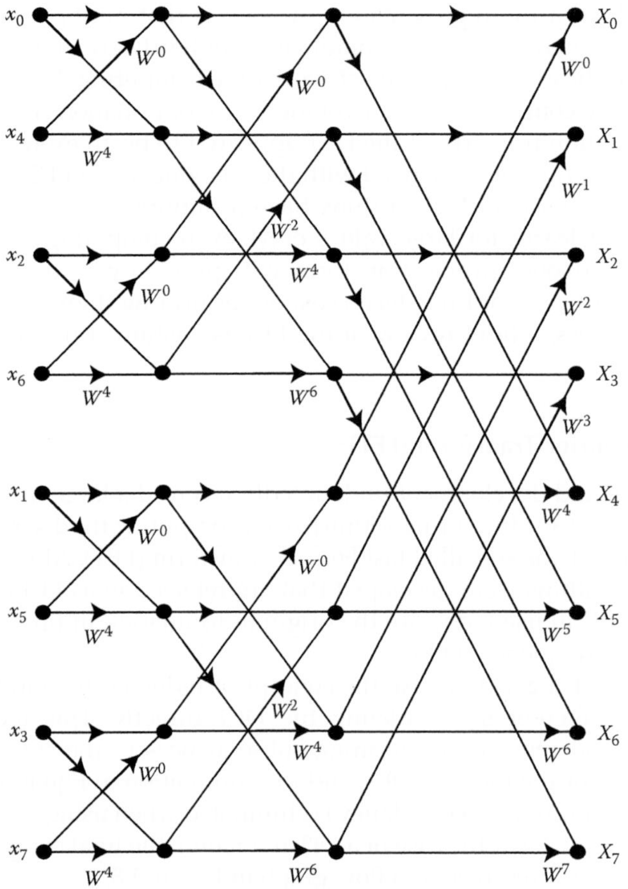

FIGURE 1.8
Flow graph for computing an 8-point DFT using a DIT FFT algorithm.

structure. A very interesting publication by S. Winograd produced a structure, based on computing DFTs with numbers of samples given by a set of small primes, that minimized the number of multiplications (though at the expense of an irregular structure and number of additions) required to compute a large sample DFT [22]. However, the exponential advances in integrated circuit technology, along with the problems of implementing irregular structures, have removed many of the reasons for minimizing the number of multiplications, particularly when considering the implementation of complex wiring nets on deep submicron integrated circuits. As discussed earlier, the DIT and DIF algorithms from the original papers are still heavily used in modern FFT hardware and software.

1.7 Arithmetic Considerations

When using fixed-point arithmetic in structures such as Figure 1.8, we need to concern ourselves with multiplicative number growth. This occurs when two or more multiplications are chained together; i.e., the output of one multiplier feeds into the input of the next in the chain. For a B-bit input word to the first multiplier in the chain, we will see a $N \cdot B$ -bit word at the output of a chain of N multipliers.

For the 8-point FFT of Figure 1.8, we find a maximum chain of three multipliers. Even if we are willing to accept an output data stream that has three times the word width of the input data stream, we can readily see, from Figure 1.8, that the inputs to adders at most of the nodes are coming from data paths with varying multiplicative chains. We clearly have to scale the inputs so that they all have the same dynamic range. We can elect to have internal nodes operating with a larger word width than input or output nodes, but that does not remove the necessity of having to scale one or both of the inputs. This scaling problem is automatically handled in floating point arithmetic units, but at the cost of considerably more hardware than required in a basic integer or fixed-point multiplier. For a fixed architecture, however, such as this 8-point DIT FFT, we know exactly where the number growth occurs, and so we can hardwire the scaling required. This is invariably much more efficient than using individual arithmetic units that automatically control the scaling (or normalization in floating point parlance) following an arithmetic operation. There are, of course, trade-offs, including longer design time and the inflexibility of a fixed architecture.

1.7.1 Arithmetic Errors in DSP

Unlike the requirements for accuracy of individual calculations that often accompany financial calculations, scientific calculations, and the like, arithmetic error in individual DSP calculations can be more appropriately classified as quantization noise in the output of a data stream process. Noise is no stranger to analog signal processing, and so its counterpart in DSP, arithmetic error, is similarly treated. In fact, the worrisome problem of a small number of pathological floating point calculations, that produce very strange errors [23], may not be of much concern with even very large errors in individual outputs of a data stream. We can think of some common data stream DSP applications such as digital audio or streaming video, where a large error in a single sample of the output of a CD recording, or in a single pixel in a streaming video application, will probably go completely unnoticed by the user.

Early DSP hardware almost exclusively used binary fixed-point or integer arithmetic. We can regard fixed point as a binary integer divided by a fixed power of 2, so that the complexity of both forms, for both multiplication and addition, is the same. The concepts of treating computational error as noise were developed assuming such number representations and associated arithmetic operations. Quantization noise can be assumed to be injected into the flow graph of a DSP algorithm. We can also assume that a noise source is injected at the input whenever there is an A/D conversion required.

Arithmetic errors are normally limited to multiplication quantization, since addition can often be performed without error by simply allowing a growth in the number of bits representing the data internal to the flow graph, based on the size of the previous addition chain at any node. Each addition in a chain will only require an increase of 1 bit to the number representation—compare this to a doubling of the word width for each multiplication in a chain. It turns out that, providing the input data to the multiplier have a relatively large dynamic range and bandwidth, the multiplier noise can be considered to be uncorrelated from sample to sample and the various noise sources (from different multipliers in a flow graph) are mutually uncorrelated random variables. This was established early on in the development of digital filter theory [21]. We can thus consider a multiplicative output noise source injection, having input data properties as listed above, as having one of three possible error probability density functions (*pdfs*), as shown in Figure 1.9.

For simplicity, we assume that the multiplier input and output have a word length of B bits with a fixed point at the left of the number so that the maximum represented value is $(1 - 1/2^B)$. As shown in figure 1.9, we assume a uniform error *pdf* between the upper and lower bounds of the quantization operation.

Given a computational flow graph, we can calculate the output noise variance using the above assumed noise model, the points of error noise injection, and the effect of the flow graph between the injection node and the output. The variance of all three of the quantization models in Figure 1.9 is the same, as shown in Equation (1.9):

$$\sigma_e^2 = \frac{1}{2^{2B} \cdot 12} \tag{1.9}$$

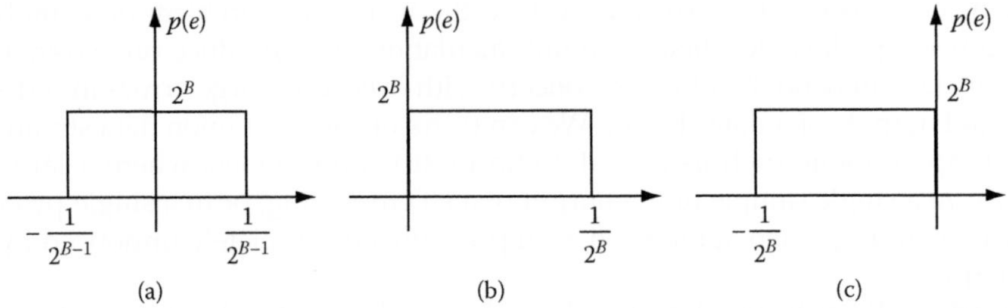

FIGURE 1.9
Probability density functions for (a) roundoff, (b) downward round, and (c) upward round.

If we assume that the graph between the injection point and the output is defined by its impulse response, $h_0, h_1, \ldots, h_{N-1}$, then the output noise, *f*, for sample *m*, f_m, is found by convolving the error noise samples with the impulse response, as shown in Equation (1.10).

$$f_m = \sum_{n=0}^{m} h_n \cdot e_{m-n} = \sum_{n=0}^{m} h_{m-n} \cdot e_n \tag{1.10}$$

Using the noise model, we can compute the variance of the output noise, as shown in Equation (1.11).

$$\sigma_{f_m}^2 = \frac{1}{12 \cdot 4^B} \sum_{n=0}^{m} h_{m-n}^2 = \frac{1}{12 \cdot 4^B} \sum_{n=0}^{m} h_n^2 \tag{1.11}$$

The steady-state value of the variance is given by Equation (1.12).

$$\sigma_f^2 = \frac{1}{12 \cdot 4^B} \sum_{n=0}^{\infty} h_n^2 \tag{1.12}$$

The mathematical tool we use to calculate the effect of quantization error noise sources is the *z*-transform [24]. The *z*-transform is the discrete system equivalent to the Laplace transform for continuous systems. It probably dates back to 1730, when DeMoivre introduced the concept of the generating function to probability theory [24]. The usefulness of the *z*-transform is that it allows difference equations to be mapped to algebraic equations. We note that in the continuous domain, the Laplace transform, in an analogous manner, allows differential equations to be mapped to algebraic equations. There is a modified *z*-transform [24] that accommodates a mix of discrete and continuous equations, but we will not need to pursue the use of that tool for this brief introduction. The *z*-transform, $H(z)$, of an infinite sequence of samples, h_n, is given by Equation (1.13).

$$H(z) = \sum_{n=0}^{\infty} h_n z^{-n} \tag{1.13}$$

where *z* is a complex variable. If the sequence $h_0, h_1, \ldots, h_{N-1}$ represents the impulse response of a discrete system, then $H(z)$ is referred to as the *transfer function* of the system. The *z*-transform, as with the Fourier transform and the Laplace transform in the continuous domain, has the bidirectional multiplication ⇔ convolution property. The DFT has the same property, but with sample indices over a finite ring, as discussed earlier. Thus, the output from a system transfer function, $H(z)$, in response to an input sequence, $X(z)$, is given by the multiplication of the transfer function with the input sequence transform, since this is the z-transform of the convolution of the input sequence with the impulse response of the system; see Equation (1.14).

$$Y(z) = X(z)\cdot H(z) \rightarrow y_m = \sum_{n=0}^{m} x_n h_{m-n} = \sum_{n=0}^{m} x_{m-n} h_n \tag{1.14}$$

We can also use the bidirectional multiplication ⇔ convolution property in the other direction. Multiplication of two input domain sequences maps to a complex convolution between their transforms, as shown in Equation (1.15):

$$\Theta(z) = \sum_{n=0}^{\infty} x_n y_n z^{-n} = \frac{1}{2\pi i}\oint Y(v) X\left(\frac{z}{v}\right) v^{-1} dv \tag{1.15}$$

where the integration is taken over a contour in the *ring of convergence* that is inside the unit circle and encloses the poles of the integrand. This result brings us back to the output noise variance computation of Equation (1.12). Instead of summing the infinite series of $\{h_n^2\}$, we can solve it in the z-transform domain, as in Equation (1.16).

$$\sum_{n=0}^{\infty} h_n^2 = \frac{1}{2\pi i}\oint H(z) H\left(\frac{1}{z}\right) z^{-1} dz \tag{1.16}$$

For the feed-forward DFT transform example of Figure 1.8, $H(z)$ is a constant, $K_{a,b}$, between any two nodes, a and b, and so $\sum_{n=0}^{\infty} h_n^2 = K_{a,b}^2$. For the complex multipliers in Figure 1.8, we need to model the real additions and multiplications between the nodes. We will use the standard computation of Equation (1.17):

$$(a+ib)\cdot(c+id) = (ac-bd) + i(ad+bc) \tag{1.17}$$

Assuming that additions (subtractions) do not inject quantization noise, we have four multiplications that can affect the noise entering the twiddle factor multiplier. For the general case, where $\{a, b, c, d\} \neq 0$, we have the flow graph shown in Figure 1.10. Clearly the noise variance gain from both the real and imaginary outputs is $(c^2 + d^2)$, which is the square of the norm of the twiddle factor. Even if we pipeline the flow graph by adding edge-triggered registers at nodes, the resulting variance gain is the same as that without the

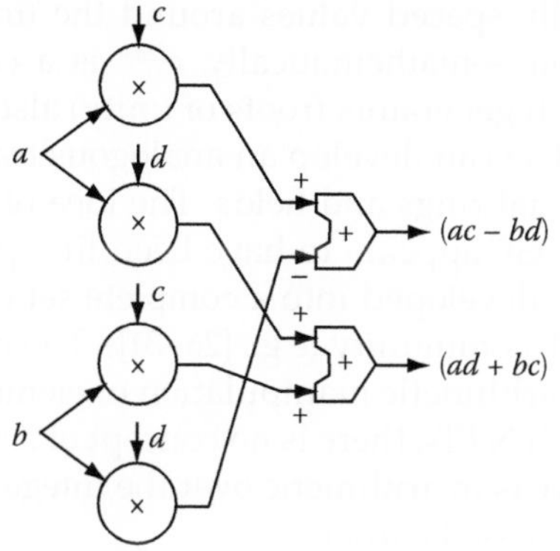

FIGURE 1.10
Structure of a twiddle factor multiplication.

pipelined registers. This is clearly correct, since the variance is measured at steady state, whereby finite delays on a graph edge are irrelevant.

1.8 Convolution Filtering with Exact Arithmetic

1.8.1 Number Theoretic Transforms (NTTs)

The convolution property of the DFT stems from the inverse DFT computation of the transforms of the sequences, $\{x\}$ and $\{y\}$, being convolved. The inverse of the transform product is shown in Equation (1.18):

$$\theta_l = \frac{1}{N}\sum_{m=0}^{N-1}\sum_{n=0}^{N-1} x_n y_m \left(\sum_{k=0}^{N-1} e^{\frac{i2\pi(l-m-n)k}{N}}\right) \tag{1.18}$$

The important property that gives rise to the expected convolution sum is that the final term on the right-hand side is zero except for $m = \langle l-n\rangle_N$. Using this property, Equation (1.18) can be simplified to the sum in Equation (1.19), which we recognize as the circular convolution sum.

$$\theta_l = \sum_{n=0}^{N-1} x_n y_{\langle l-n\rangle_N} = \sum_{n=0}^{N-1} y_n x_{\langle l-n\rangle_N} \tag{1.19}$$

The property we exploit here is based on the circular nature of the complex exponential function, $e^{i2\pi/n}$, that generates (by repeated multiplications) equally spaced values around the unit circle that repeat every N multiplications—mathematically, $e^{i2\pi/n}$ is a complex Nth root of unity. We can find such generators (roots of unity) also within finite mathematical systems, and so can develop an analogous transform structure to the DFT, but over special rings and fields. The idea of exploiting a *number theoretic transform* in DSP appears to have been first published by Pollard [25], and then quickly developed into a complete set of fast algorithms by researchers in the DSP community (e.g., [26–31]). The notion of exact arithmetic stems from the arithmetic manipulation of elements of the rings and fields that support such NTTs; there is no concept of a result being approximately correct, as there is in arithmetic over the integers with quantization errors, as we have discussed earlier.

1.8.2 An NTT Example

For completeness the following example is used to demonstrate the concept. We define a finite field using the integers 0, 1, 2, … , 16. Standard addition and multiplication are performed followed by a reduction, Mod 17. This field will be referred to as GF(17), a Galois field. A field has two operators: in this case, $\oplus_{17} \Rightarrow \langle a+b\rangle_{17}$ and $\otimes_{17} \Rightarrow \langle a\times b\rangle_{17}$. For subtraction we define an additive inverse of a, $\bar{a} = 17 - a$; note that 17 maps to 0 when applying the modular operation. A field is a ring where all of the elements also have a multiplicative inverse, $\hat{a}$, where $\langle \hat{a}\times a\rangle_{17} = 1$. Thus additive inverses are elements within the ring that can be treated as negative numbers for subtraction, and multiplicative inverses are elements within the ring that can be treated as reciprocals for exact division (approximate division, unlike in ordinary arithmetic, has no meaning over such finite rings or fields). GF(17) also has a generator, $g = 3$, that will generate all of the field elements, except 0, by repeated multiplications.

Table 1.1 shows the elements, x, additive inverses, $\bar{x}$, multiplicative inverses, $\hat{x}$, and generated elements of the finite field, GF(17). The generated elements are given by $g^x = 3^x$. Thus the ordered field elements, x, are the exponents, or indices, of the generated elements. Note that the generated elements repeat after 16 multiplications. This is analogous to the way that $e^{i2\pi/n}$ generates repeated values around the unit circle in the complex z-plane, and so the

TABLE 1.1

Finite Field Elements: GF(17)

x	0	1	2	3	4	5	6	7	8	9	10	11	12	13	14	15	16
$\bar{x}$	0	16	15	14	13	12	11	10	9	8	7	6	5	4	3	2	1
$\hat{x}$	–	1	9	6	13	7	3	5	15	2	12	14	10	4	11	8	16
$g = 3$	–	3	9	10	13	5	15	11	16	14	8	7	4	12	2	6	1

generator, g, is the 16th *root of unity* we need to form a 16-sample NTT finite field transform. This transform and its inverse are shown in Equation (1.20).

$$
\begin{aligned}
X_k &= \left\langle \sum_{l=0}^{15} x_l 3^{lk} \right\rangle_{17} \\
x_l &= \left\langle 16 \cdot \sum_{k=0}^{15} X_k 6^{lk} \right\rangle_{17}
\end{aligned}
\tag{1.20}
$$

The multiplicative inverse of 16, $\hat{16} \equiv 16$, is an element of the ring, and we can show that all values of $\sum_{k=0}^{15} X_k g^{kl}$ are multiples of 16 over the field. Therefore the two conditions for using this multiplicative inverse are met. Using the negative representation, $16 \equiv -1$, then, from above, $\widehat{16} \equiv -1$ or $16 \otimes_{17} \widehat{16} \equiv -1 \otimes_{17} -1 = 1$, which provides an example of using negative representations to perform calculations over the field. The indices of the generator can also be considered to be finite field logarithms over GF(17), and thus can be used to simplify multiplication to addition, as ordinary logarithms are used over the real numbers (see our earlier discussions). Additions over GF(17) can be simplified because the form of the modulus, 17, is $M = 2^4 + 1$, and so we can implement addition with a 4-bit adder and a small amount of extra logic [29]. If we choose the form $M = 2^{2^t} + 1$, this is known as a Fermat number and is prime up to $t = 4$ [32]. Note that the index calculus computes over a ring with modulus 2^{2^t}, and so ordinary binary adders can be used to perform multiplication.

Our example NTT is, in fact, a Fermat NTT (FNTT), although we have used such a small Fermat prime that the usefulness of larger prime FNTTs is somewhat hidden by the obvious connection of the sample length to the dynamic range (they are the same in this example). The first NTTs seriously explored for convolution applications were FNTTs [26]; although there are always links between word length and sample size, there was enough impetus to solve the large hardware cost of implementing FFTs in the 1970s. Considerable international research effort was expended in exploring all aspects of number theoretic techniques for at least two decades (e.g., [30,31,33]).

1.8.3 Connections to Binary Arithmetic

We make use of additive inverses when using two's complement binary arithmetic. For example, an 8-bit binary number can be regarded as a finite ring with 256 elements and ring operators of ordinary addition and multiplication followed by a Mod 256 reduction. This is a trivial reduction operation to implement in hardware; we simply ignore overflow carries past the most significant bit (MSB), and ignoring the carries requires no hardware! The term *taking the two's complement* is actually the additive inverse computation of Equation (1.21):

$$\bar{x} = 256 - x = (255 - x) + 1 \tag{1.21}$$

We note that (225 – x) is the one's complement of x (simply invert each of the eight binary digits). The well-known two-step procedure is thus: (1) invert the bits, and (2) add 1. Subtraction, $x - y$, is now carried out by adding x to the two's complement of y, and ignoring any carries out of the eighth place. For convenience, and to maintain an association with signed integers, we often refer to the lower half of the ring elements, {0 … 127}, as positive numbers in the ring and the upper half, {128 … 255}, as negative numbers. An interesting situation occurs if we know that the final result, independent of the number of additions (subtractions), is within the dynamic range of the ring; the result will be correct even if there are overflows during the computation. The reason is that our ring operations, of normal addition followed by a reduction operation, carried out over a sequence of additions (subtractions), yield the same value as performing normal addition (subtraction) for the entire sequence followed by a single reduction operation. This means that we can reduce the amount of addition hardware (reduce the word length) if we know, a priori, that the result will be within a given smaller dynamic range than required for the intermediate computations. The rub here is in finding an application where a sequence of additions (subtractions) yields such a smaller dynamic range without knowing specifically what the intermediate computations will be—basically having some other means of determining the dynamic range. Such applications are difficult to find, but we show, in Chapter 5, that one of our architectures yields such a property, and this can be used to reduce the word length in adders and read-only memory (ROM) output word length.

Although the 8-bit binary system is only a ring, some of the elements have multiplicative inverses; in fact, all the elements that are relatively prime to the modulus, 256, which are all of the odd numbers, have multiplicative inverses. We can carry out division by an odd element of the ring using the multiplicative inverse of that element. The only proviso is that the quotient exists as an element of the ring; i.e., it is an integer of < 256. As an example, divide 21 by 3, using the multiplicative inverse of 3, which is 11. Although 21 is actually greater than the dynamic range, 16, of the 4-bit binary number, we know that the result, 7, is within the positive part of the dynamic range, and so the calculation will yield the correct result. In fact, we only need to perform the arithmetic using 4-bit numbers, even though intermediate results may be outside the dynamic range.

We can thus use the residue of 21 Mod 16 (5) instead of 21, and the calculation becomes 5 × 11 = 55 with a final reduction Mod 16, to give the correct answer of 7. The reduction Mod 16 is implemented by simply ignoring any carries out of the 4th place using a binary multiplication algorithm—so it is essentially free!

This seems quite a powerful result, yet there is an Achilles' heel: we have to have a priori knowledge that the answer is an integer within the range of our number system, and it is difficult to think of any general practical problem for which this is guaranteed. We see that the NTT calculation requires such a division, but this is a specific case where we know that the division is exact and the quotient is within the range of our number system. We will find a more general example of this effect when we discuss an index calculus processor for a two-dimensional logarithmic number system (2DLNS) correlator in Chapter 5.

1.9 Summary

Unlike many books on computer arithmetic, which target general computer applications, in this work we target specific applications. DSP has been a driver for many applications of alternative number representations and, not surprisingly, has been a target for the application of the multiple-base representations discussed in this book. As such we have spent a few pages in this introductory chapter on the basics of DSP in order to discuss some of the special arithmetic aspects. This has included data stream processing, sampling, and early examples of DSP flow graphs using the discrete Fourier transform, and the first fast algorithms. We have also discussed the representation of arithmetic errors as digital noise, rather than characterizing individual arithmetic errors resulting from the use of finite precision arithmetic. We have also detailed the place of finite ring (exact) arithmetic, in special DSP algorithms, and the link to the well-known two's complement binary representation.

Both analog and digital computational aids have been used in the past, and we include both in the brief history at the beginning of the chapter. Clearly, what was the domain of analog circuits in many applications in the past is now firmly in the grasp of digital circuits and systems. However, our inclusion of analog computation in this chapter is germane to some of our work on using symbolic substitution rules to use analog circuits in a digital medium. We discuss such approaches in Chapters 2 and 3. In Chapter 3 we use analog circuits in a cellular neural network, to implement a set of substitution rules.

We have also briefly touched on logarithmic arithmetic with a discussion of slide rules—a portable calculator that was relegated to the "nostalgia storage bin" soon after the first portable electronic calculators appeared in the early 1970s. Even so, a trick used by aficianados of the slide rule has an important role to play in modern logarithmic arithmetic, and we will see this slide rule trick in operation with the discussions of multidimensional logarithmic arithmetic starting in Chapter 5.

References

1. Morrison, P., Morrison, P. Commentaries: Wonders—The Physics of Binary Numbers. *Scientific American Magazine* 274, no. 2 (1996).
2. Pullan, J.M. *The History of the Abacus*. New York: Frederic A. Praeger, 1969.
3. Freeth, T., Jones, A., Steele, J.M., Bitsakis, Y. Calendars with Olympiad Display and Eclipse Prediction on the Antikythera Mechanism. *Nature* 454 (2008): 614–617.
4. Ehret, T. Old Brass Brains: Mechanical Prediction of Tides. *ACMS Bulletin* (2008): 41–44.
5. Swade, D. *The Difference Engine*. New York: Penguin USA, 2002.
6. Wilkes, M.V. Babbage as a Computer Pioneer. *Historia Mathematica* 4, no. 4 (1977): 415–440.
7. Abramowitz, M., Stegun, I.A. *Handbook of Mathematical Functions*. Washington, DC: National Bureau of Standards, 1968.
8. Kingsbury, N.G., Rayner, P.J.W. Digital Filtering Using Logarithmic Arithmetic. *Electronics Letters* 7 (1971): 56–58.
9. Swartzlander, E.E., Alexopoulos, A.G. The Sign/Logarithm Number System. *IEEE Transactions on Computers* 37 (1975): 1238–1242.
10. Taylor, F.J., Gill, R., Joseph, J., Radke, J. A 20 Bit Logarithmic Number System Processor. *IEEE Transactions on Computers* 37, no. 8 (1988): 190–200.
11. Coleman, J.N., Chester, E.I., Softley, C.I., Kadlec, J. Arithmetic on the European Logarithmic Microprocessor. *IEEE Transactions on Computers* 49, no. 7 (2000): 702–715.
12. McCulloch, W.S., Pitts, W.H. A Logical Calculus of the Ideas Immanent in Nervous Activity. *Bulletin of Mathematical Biophysics* 5 (1943): 115–133.
13. Moore, G.E. Cramming More Components onto Integrated Circuits. *Electronics*, 19, no. 4 (1965).
14. Faggin, F. The Making of the First Microprocessor. *IEEE Solid-State Circuits Magazine* (2009): 8–21.
15. Jullien, G.A. Implementation of Multiplication, Modulo a Prime Number, with Applications to Number Theoretic Transforms. *IEEE Transactions on Computers* C-29, no. 10 (1980): 899–905.
16. International Technology Roadmap for Semiconductors. ITRS 2009 Edition. Online documents, International Technology Roadmap for Semiconductors, 2009.
17. Bracewell, R.N. *The Fourier Transform and Its Applications*. 2nd ed. New York: McGraw-Hill, 1978.
18. Dirac, P. *Principles of Quantum Mechanics*. 4th ed. Oxford: Clarendon Press, 1958.
19. Burrus, C.S., Parks, T.W. *DFT/FFT and Convolution Algorithms*. New York: John Wiley & Sons, 1985.
20. Cooley, J.W., Tukey, J.W. An Algorithm for the Machine Calculation of Complex Fourier Series. *Mathematics of Computation* 19 (1965): 297–301.
21. Gold, B., Rader, C.M. *Digital Processing of Signals*. New York: McGraw-Hill, 1969.
22. Winograd, S. On Computing the Discrete Fourier Transform. *Mathematics of Computation* 32 (1978): 175–199.

23. Muller, J.-M. *Elementary Functions: Algorithms and Implementation*. Boston: Birkhåuser, 1997.
24. Jury, E.I. *Theory and Application of the z-Transform Method*. New York: John Wiley & Sons, 1964.
25. Pollard, J.M. The Fast Fourier Transform in a Finite Field. *Mathematics of Computation* 25 (1971): 365–374.
26. Agarwal, R.C., Burrus, C.S. Fast Convolution Using Fermat Number Transforms with Applications to Digital Filtering. *IEEE Transactions on Acoustics, Speech and Signal Processing* ASSP-22 (1974): 87–97.
27. Reed, I.S., Truong, T.K. The Use of Finite Fields to Compute Convolution. *IEEE Transactions on Information Theory* IT-21 (1975): 208–213.
28. Nussbaumer, H.J. Digital Filtering Using Complex Mersenne Transforms. *IBM Journal of Research and Development* 20 (1976): 498–504.
29. Leibowitz, L.M. A Simplified Binary Arithmetic for the Fermat Number Transform. *IEEE Transactions on Acoustics, Speech and Signal Processing* ASSP-24 (1976): 216–225.
30. McClellan J.H., Rader, C.M. *Number Theory in Digital Signal Processing*. Englewood Cliffs, NJ: Prentice-Hall, 1979.
31. Blahut, R.E. *Fast Algorithms for Digital Signal Processing*. Reading, MA: Addison-Wesley, 1985.
32. Robinson, R.M. Mersenne and Fermat Numbers. *Proceedings of the American Mathematics Society* 5 (1954): 842–846.
33. Jullien, G.A. Number Theoretic Techniques in Digital Signal Processing. In *Advances in Electronics and Electron Physics*, Vol. 80, 69–163. New York: Academic Press, 1991.

2

The Double-Base Number System (DBNS)

2.1 Introduction

In this chapter we formally introduce the DBNS. The representation was discovered by one of the authors (Dimitrov) and a colleague, T.V. Cooklev, when working on computation problems in graph theory [1] and cryptography [2]. Dimitrov introduced the technique to our research group (VLSI Research Group) at the University of Windsor in Ontario, Canada, in 1995. The main interest of the group was in digital signal processing (DSP) at that time, and so we started to look at the application of DBNS representations to filter architectures. The earliest publication from the VLSI group was from a presentation given at the SPIE annual conference in 1996 [3]. One of our ideas in that paper was to use cellular neural networks to implement reduction rules for DBNS arithmetic, a topic we have explored for a number of years [4], and we discuss this approach in some detail in Chapter 3.

A major outcome from our work is the use of an index calculus to compute with multiple bases forming a logarithmic type representation, which we have termed the multidimensional logarithmic number system (MDLNS). This has turned out to provide some very efficient implementations for specific DSP architectures, and we devote Chapters 5 through 9 to explore our findings in the use and, sometimes peculiar, properties of the MDLNS.

Over the past 15 years, we have explored the use of representations using two or more bases: for performing basic arithmetic operations [5–7]; for applications in DSP [5,8–10], with many integrated circuit designs—some of which are provided in Chapter 9; for other computationally exacting algorithms in computer graphics and the like; and for cryptography [11–15], where exact computations with very large integers form the computational backbone of current security systems.

2.2 Motivation

In many applications the computational complexity of algorithms crucially depends upon the number of zeros of the input data in the corresponding number system [16–20]. Number systems are often chosen to enable a reduction of the complexity of the arithmetic operations; the most popular are, perhaps, signed digit number systems [20]. An analysis of the expected number of zeros in the representation of arbitrary integers in the binary signed digit number system shows that on average, for long word lengths, 33% fewer adders are needed to perform multiplication than binary [21,22]. In these number systems we need, on average, $O(\log N)$ nonzero digits to represent the integer N. Although redundancy appears to be an inefficiency in performing arithmetic, it turns out that signed digit arithmetic provides hidden efficiencies. We can certainly trace the use of signed binary in computer arithmetic back to the classical paper by Booth [23], in which two's complement addition operations used in performing multiplication with signed digits are controlled by subsequences of adjacent digits in the multiplier, with no addition operations having to be performed for subsequences of the same digits. For computation technologies, where shifting is far less complex than addition (mostly), this technique introduces savings in hardware and power consumption, and provides increases in speed over the conventional shift-and-add algorithm using standard binary representations. Many other major advantages have been found for redundant signed digit binary arithmetic, and it is our thesis that the investigation of other redundant representations may uncover additional advantages, particularly in the case of application-specific computational hardware. To this end, we will now formally introduce the DBNS, from foundations to algorithms.

2.3 The Double-Base Number System

2.3.1 The Original Unsigned Digit DBNS

We formally define a number system, allowing as digits only 0,1 and requiring $O(\log N)$ nonzero digits, as the double-base number system (DBNS), using bases 2 and 3, that is, a representation having the form of Equation (2.1).

$$x = \sum_{i,j} d_{i,j} 2^i 3^j \tag{2.1}$$

Clearly, by setting $j = 0$, the binary number system becomes a special case (and valid member) of the above representation, and for this introduction we consider the classical example of using signed digit binary to efficiently represent the number $x = 127$. In standard 8-bit binary, where $x = \sum_{i=0}^{7} x_i 2^i$, $127 \rightarrow 01111111$. Since arithmetic operations with this number representation have a hardware cost that increases with the number of nonzero digits (e.g., multiplication), 127 can be considered to be very inefficiently represented using standard binary, since 7 out of the 8 digits are 1. However, allowing the digits to take on both positive and negative values, i.e., $x_i \in \{-1,0,1\}$, we can rewrite $127 = 128 - 1 \rightarrow 1000000\bar{1}$, where $\bar{1} \equiv -1$. Clearly this representation is redundant since we note that standard binary is a valid member of signed binary, and so we have two signed digit representations for 127. Such redundancy introduces inefficiencies, but this is overweighed by the considerable reduction in arithmetic complexity afforded by the large reduction in nonzero digits [24].

2.3.2 Unsigned Digit DBNS Tabular Form

If we now display the digits for the unsigned DBNS representation over orthogonal bases 2 and 3, we obtain a two-dimensional table with the two dimensions defined by i and j from Equation (2.1). To demonstrate the considerable redundancy in the DBNS, three representations for $x = 127$ are shown in Figure 2.1; each of them has the same number of nonzero digits: three. It is a fairly simple task to produce other representations (there are a total of 783 possible representations for $x = 127$), such as those in Figure 2.2; however, these representations require more than three nonzero digits. It turns out that for $x = 127$, the minimum number of nonzero digits is three; we

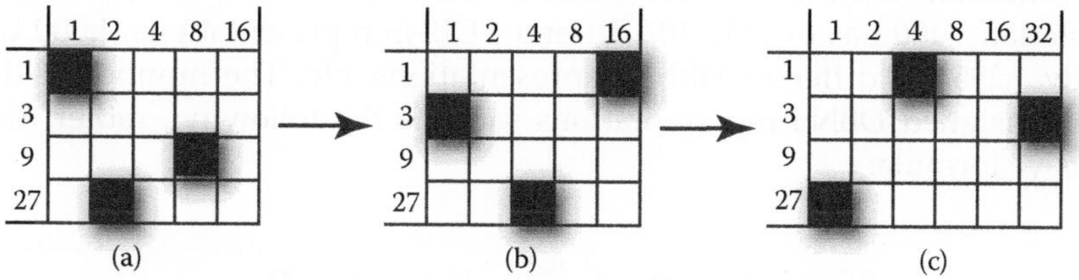

FIGURE 2.1
Three canonic DBNS representations for $x = 127$.

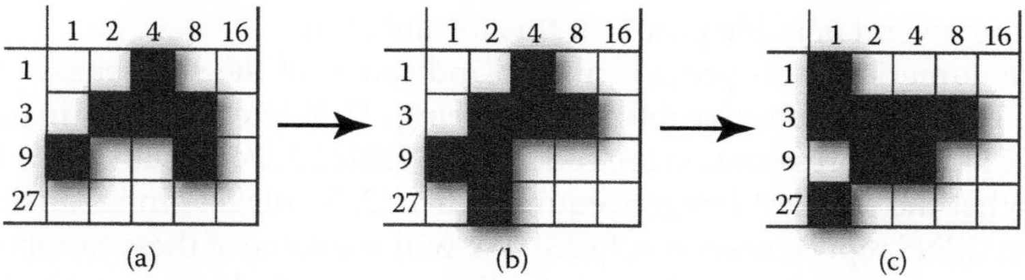

FIGURE 2.2
Three noncanonic representations for $x = 127$.

will refer to such a minimum representation as canonic, and representations with larger than the minimum number of nonzero digits as noncanonic. If we are close to a minimum representation, we will refer to it as near canonic (say within a digit or two of the canonic representation).

2.3.3 The General Signed Digit DBNS

The general signed digit DBNS is defined as below:

Definition 2.1 (DBNS): Given p, q, two different prime numbers, the double-base number system (DBNS) is a representation scheme in which every positive integer x is represented as the sum or difference of $\{p, q\}$-integers, i.e.:

$$x = \sum_{i=1}^{l} s_i p^{a_i} q^{b_i} \text{ with } s_i \in \{-1, 1\} \text{ and } a_i, b_i \geq 1 \tag{2.2}$$

We note that the DBNS is quite redundant (and sparse) without having to resort to using signed digits (i.e., with $s_i = 1$). In fact, it is much more redundant than signed digit binary. As with signed digit binary, we accept the redundancy in exchange for extra degrees of freedom that are able to yield efficiencies (which will be exposed as we progress through this and further chapters).

2.3.4 Some Numerical Facts

Relating to the sparseness of the DBNS, if we assume unsigned double-base representations only, we can show that 10 has exactly 5 different DBNS representations, 100 has exactly 402 different DBNS representations, 1,000 has exactly 1,295,579 different DBNS representations, etc. The number of different unsigned DBNS representations satisfies the following, rather nice, recursive formula:

$$\begin{aligned} f(n) &= f(n-1) + f\left(\frac{n}{3}\right) \qquad \text{if } n = (mod\ 3), \\ &= f(n-1) \qquad \text{otherwise.} \end{aligned}$$

There are several possible proofs of this formula [14].

Some numerical facts provide a good indication of the sparseness of the DBNS. The smallest positive integers requiring l $\{2, 3\}$-integers in its unsigned canonic DBNS representation are collected in Table 2.1. For instance, the table shows that the smallest integer requiring five $\{2, 3\}$-integers in its unsigned canonic DBNS representation is 18,431. We will use some of these numbers to prove very nontrivial upper bounds for the number of additions sufficient to multiply by any k-bit number. All of the numbers in Table 2.1 can be found, at

TABLE 2.1
Smallest Value of x Requiring l {2,3}-Integers in Its Canonic DBNS Representation

l	x
1	1
2	5
3	23
4	431
5	18,431
6	3,778,433
7	1,441,896,119

least in principle, by using exhaustive search. More efficient searching methods are possible, but they are beyond the scope of this book.

If one considers signed representations, then the theoretical difficulties in establishing the properties of this number system drastically increase. To wit, it is possible to prove that the smallest number that cannot be represented as the sum or difference of two {2, 3}-integers is 103. The next limit is 4,985, but to prove it rigorously, one has to prove the following: the Diophantine equations $\pm 2^a3^b \pm 2^c3^d \pm 2^e3^f = 4985$ do not have solutions in integers.

The proof that these equations do not have solutions in integers has been recently found by Dimitrov and Howe [25]. More to the point, it is now known that the smallest integer that cannot be represented as the sum or difference of four {2, 3}-integers is 641,687, and the smallest integer that cannot be represented as the sum or difference of five {2, 3}-integers is 326,527,783.

2.4 The Greedy Algorithm

Comparing the noncanonic representations of Figure 2.2 with the canonic representations of Figure 2.1, we see a tendency for the digits to "clump" around the left and upper sides of the two-dimensional tables. A more striking example is shown in Figure 2.3 for $x = 54$, which has 93 possible representations. In this example, the canonic representation (Figure 2.3(b)) needs only one digit. We can find very inefficient representations by substituting higher-value digit weights with two or more lower-value digit weights. A canonic representation needs to use the opposite strategy. Although finding canonic representations turns out to be a very hard problem for large integers, a near-canonic form (near-canonic double-base representation

(NCDBNR)) can be found using a "greedy" algorithm, where successive steps in the algorithm produce a monotonically decreasing sequence of digit weights.

2.4.1 Details of the Greedy Algorithm

The greedy algorithm pseudocode is shown in Figure 2.4, and from Table 2.2 we see that the algorithm generates the leftmost canonic representation of $x = 127$ in Figure 2.1. Although, in this example, we see that the greedy algorithm produces a canonic representation, for most values of x the algorithm will not yield a canonic form. We can see this using $x = 41$ in Figure 2.5, where the canonic representation (Figure 2.5(b)) requires only two digits but the greedy algorithm produces a three-digit representation (Figure 2.5(a)). Although there are problems with applying the greedy algorithm, such as finding the maximum digit weights to subtract at each step of the algorithm, solutions are being published [25]. The most important feature of the greedy algorithm is that it guarantees an expansion of sublinear length. We provide the following theorem [5], which demonstrates this striking feature of the algorithm, and we also provide the proof, for completeness.

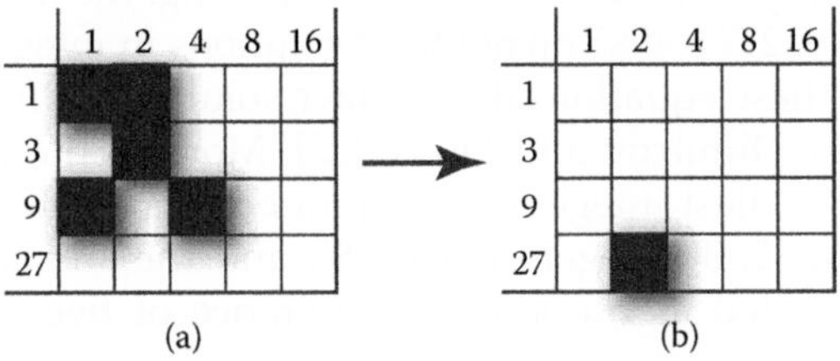

FIGURE 2.3
Two representations for $x = 54$.

Input: Positive integer x;
Output: 2-integers a_i, such that $\Sigma a_i = x$.

```
procedure       greedy (x)
{
        if (x > 0) then do
        {
                find the largest 2-integer
                w < x;
                write (w);
                x: = x - w;
                greedy (x);
        }
        else exit;
}
```

FIGURE 2.4
Greedy algorithm (pseudo code).

TABLE 2.2

Steps in the Greedy Algorithm for the DBNS Representation of 127

x	w	$x - w$
127	$3^3 \cdot 2^2 = 108$	19
19	$3^2 \cdot 2^1 = 18$	1
1	$3^0 \cdot 2^0 = 1$	0

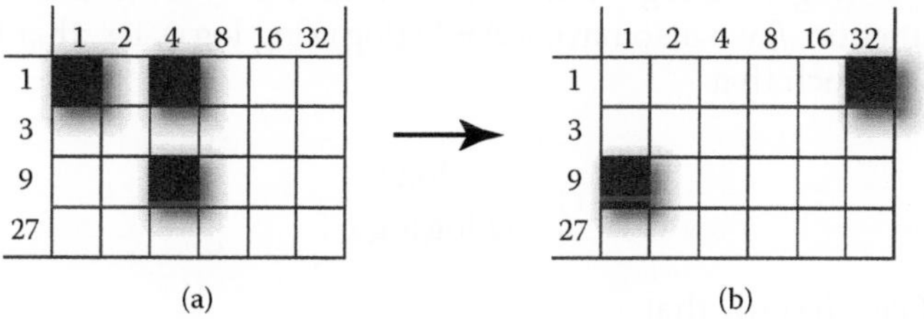

FIGURE 2.5
Greedy vs. canonic representations for $x = 41$.

Theorem 2.1: The greedy algorithm terminates after $k = O$ (log x/log log x) steps.

Proof: First we mention that $k = O$ (log x), simply by taking the 2-adic and 3-adic expansion of x. R. Tijdeman [26] has shown that there exists an absolute constant $C > 0$ such that there is always a number of the form 2^a3^b between $x–x/(\log x)^C$ and x. Now we put $n_0 = x$, and the Tijdeman result implies that there exists a sequence:

$$n_0 > n_1 > n_2 \ldots > n_l > n_{l+1} \tag{2.3}$$

such that $n_i = 2^{a_i}3^{b_i} + n_{i+1}$ and $n_{i+1} =< n_i/(\log n_i)^C$ for $i = 0, 1, 2, \ldots l, \ldots$ Obviously the sequence of integers, $\{n_i\}$, obtained via the greedy algorithm, satisfies these conditions. Here we choose $l = l\ (x)$ so that

$$n_{l+1} \leq f(x) < n_l \tag{2.4}$$

for some function f to be chosen later.

It follows that we now can write a member of the sequence as a sum of k terms of the form 2^a3^b, where $k = l\ (x) + O$ (log $f(x)$). We now have to estimate $l(x)$ in terms of $f(x)$, and to choose an optimal $f(x)$. Note that if $i < l$, then $n_i > n_l > f(x)$; hence $n_{i+1} < n_i/(\log n_i) < n_i/(\log f(x))^C$. This implies the following inequality:

$$f(x) < n_l < \frac{x}{(\log f(x))^{lC}} \tag{2.5}$$

Thus we find $l(x) < \frac{\log x - \log f(x)}{C \log\log f(x)}$. We now take

$$f(x) = \exp \frac{\log x}{\log\log x} \tag{2.6}$$

The function in Equation (2.6) is the largest possible (apart from constant) to which we can show that any number on the interval [1, $f(x)$] can be written as a sum of $k = O$ (log$x$/log log$x$) terms of the form 2^a3^b. We want to show that with this function, f, we also have $l(x) = O$ (log$x$/log log$x$), i.e., that there is a constant $D > 0$ such that

$$l(x) < \frac{1}{D} \frac{\log x}{\log\log x} \tag{2.7}$$

Thus it suffices to show that

$$\frac{\log x - \dfrac{\log x}{\log\log x}}{C \log \dfrac{\log x}{\log\log x}} < \frac{1}{D} \frac{\log x}{\log\log x} \tag{2.8}$$

This inequality can be rewritten as

$$D \log\log x + C \log\log\log x < C \log\log x + D \tag{2.9}$$

which is true if $D < C$ and x large enough. Such a D exists, and so the proof is complete. ■

2.4.2 Features of the Greedy Algorithm

In order to assess the application of the greedy algorithm to producing near-canonic representations, an experiment was performed with 1,000 randomly selected 215-bit integers. We predict (from Theorem 2.1) that the expected number of nonzero digits is 27.75; the occurrence of the number of 2-integers peaks at about 30, as shown in Figure 2.6, which successfully demonstrates the efficacy of our algorithm. A variety of computational experiments show that the largest 2-integer, smaller than x, occurs in at least one of the CDBNRs of x in about 80% of the cases. This observation, along with Theorem 2.1, allows an estimate that the greedy algorithm returns a CDBNR with probability $0.8^{\log x/\log\log x}$; fortunately, this tends to zero very slowly [5].

We also find that the greedy algorithm produces a representation requiring on average O (log$x$/log log$x$) 2-integers. This NCDBNR provides a sufficiently sparse representation of x to make it very useful.

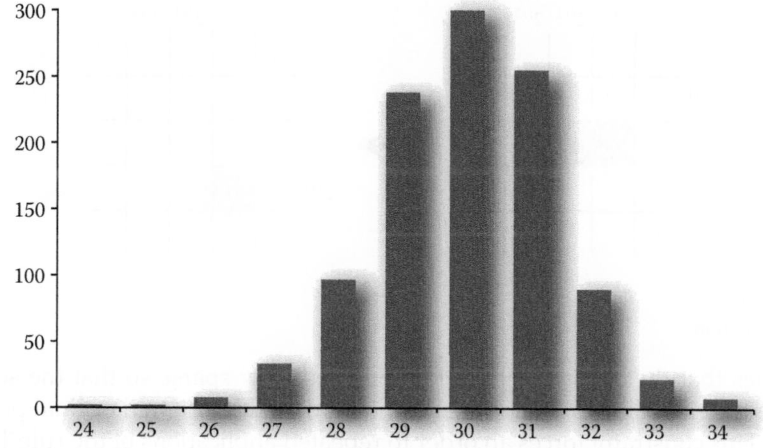

FIGURE 2.6
Occurrence of 2-integers for 1,000 randomly chosen 215-bit integers [5,10].

2.5 Reduction Rules in the DBNS

The NCDBNR would seem to be a sufficiently sparse DBNS representation, but the sparseness soon disappears once we start to manipulate pairs of numbers when performing arithmetic of addition and multiplication. In fact, we need to address not just the sparseness but also the efficiency of the representation prior to performing arithmetic operations. It is clear that if numbers are represented in as sparse a form as possible, then any manipulations with these representations will be less complex than if we choose a representation with many nonzero digits. In Section 2.3.1 we used signed digit binary to demonstrate this property. Because of the two-dimensional structure of the DBNS representation, we will develop some simple substitution rules that allow an increase in the sparsity of the representation. These rules will also be used to transform representations into forms that are very efficient for performing addition, a subject that will be dealt with in detail in Chapter 3.

2.5.1 Basic Reduction Rules

In this section we discuss two simple rules that are used to reduce noncanonic representations to near canonic. The first rule we refer to as *row reduction*, as shown, using the two-dimensional table representation, in Figure 2.7; the rule can be stated mathematically as in Equation (2.10).

$$2^i3^j + 2^{i+1}3^j = 2^i3^{j+1} \tag{2.10}$$

The meaning of the term *reduction* is self-evident, and this rule provides a 2:1 reduction in digits for which the rule applies. Application of the rule

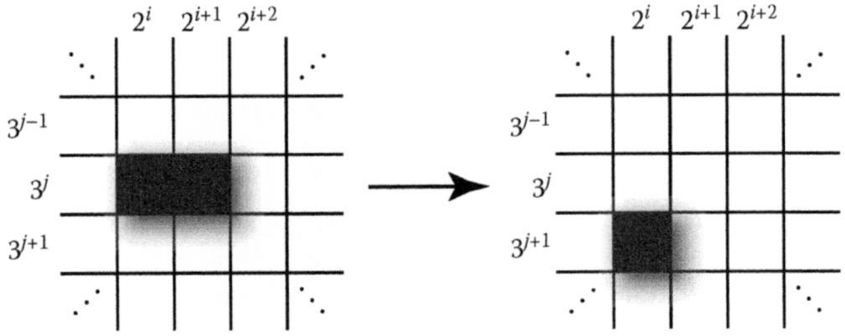

FIGURE 2.7
Row reduction.

assumes that the original representation is suitably sparse so that the substituted digit positions do not interfere with the other original or final positions. This constraint forces an order to repeated applications of the rule, but such an order is not necessarily unique.

A similar rule, which we refer to as *column reduction*, is shown in Figure 2.8 and can be stated mathematically as

$$2^i3^j + 2^i3^{j+1} = 2^{i+2}3^j \tag{2.11}$$

The same restrictions apply to the application of the column reduction rule as to the row reduction rule.

2.5.2 Applying the Basic Reduction Rules

We provide several examples of the application of the basic reduction rules, taken from previous examples of DBNS representations.

Figure 2.9 shows the application of both row and column reduction rules to reducing the noncanonic representation of $x = 127$ given in Figure 2.2(c). The selected input digits to the reduction rule are shown in gray for each step of the reduction process. In Figure 2.9(a) we are able to apply both a column and row reduction rule in parallel, with the result shown in Figure 2.9(b). We then again apply both row and column rules in Figure 2.9(b), and a single row reduction rule in Figure 2.9(c), with the final result shown in Figure 2.9(d). Interestingly this final result is the same canonic representation as in Figure 2.1(a), though, as with the greedy algorithm, there is no guarantee of obtaining canonic representations from the application of reduction rules.

To demonstrate the nonuniqueness of the application of reduction rules, we apply a different order of reduction rules to the same example, with the results shown in Figure 2.10. In this case we apply two column reduction rules in Figure 2.10(a): a single row reduction rule in Figure 2.10(b), and a column reduction rule in each of Figure 2.10(c) and (d). The final result in Figure 2.10(e) is a canonic representation, in this case, the same as that in Figure 2.1(c). In Figure 2.11 there is a demonstration of only applying a single rule, in this case row reduction. The final result is near canonic, not canonic, but has an interesting property that can be used in addition and multiplication,

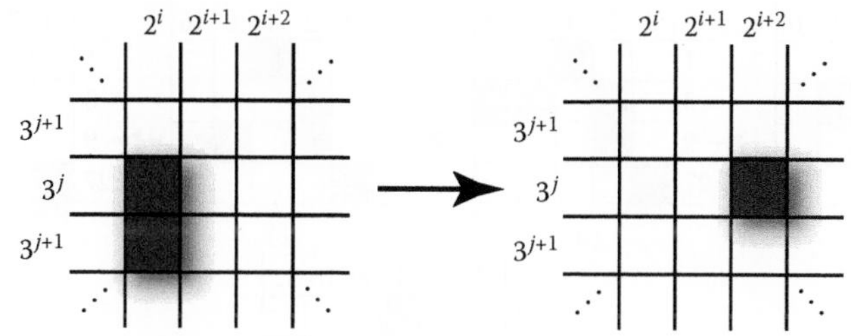

FIGURE 2.8
Column reduction.

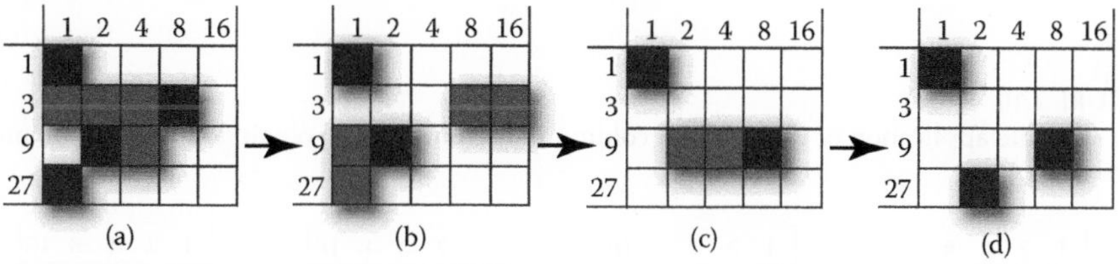

FIGURE 2.9
Row and column reduction applied two times in parallel and one extra row reduction to obtain a canonic form for $x = 127$.

which is discussed in Chapter 3. We note from these three examples of DBNS reduction that there is not a fixed number of iterations for a given number of digits at the input since we do not know the final number of digits, and hence the total reduction of digits, nor the number of parallel reductions that can take place in any of the iterations.

So far we have carefully selected examples where there are no overlapping digits, that is, where the result of applying a reduction rule results in the output digit from the application of the rule overlapping with an existing digit that is not being removed in that iteration. We can see the effect of such an overlap in Figure 2.12, where the noncanonic representation of $x = 41$ from Figure 2.3(a) is reduced down to the canonic form from Figure 2.3(b). In Figure 2.12(a) we apply the row reduction rule to the digits shown in gray. There is no conflict in applying this rule, since the output digit from the rule appears in a previously empty position in the table. However, a second application of the rule to Figure 2.12(b) results in an output digit that conflicts with a digit position from the previous iteration. We show this as two overlapping digits in Figure 2.12(c). Since the two digits together represent one of the digits multiplied by 2, we can simply replace the overlapping digits by a digit one place to the right of the overlapped digit position. Fortunately, in this example, the location to the right of the overlapped digit position is empty. If we recall the example from Figure 2.11, the final result, which had empty positions to the right of all digits in the representation, would seem ideal for applications of conflicting digits. For reference, we will

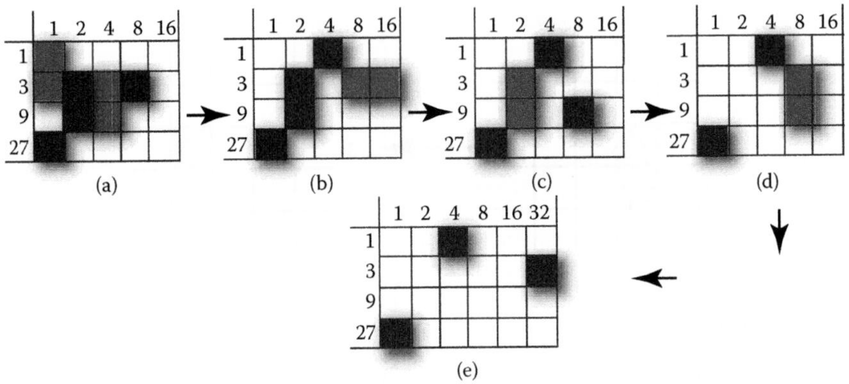

FIGURE 2.10
An alternate application of the row and column reduction rules to obtain a different canonic form for $x = 127$.

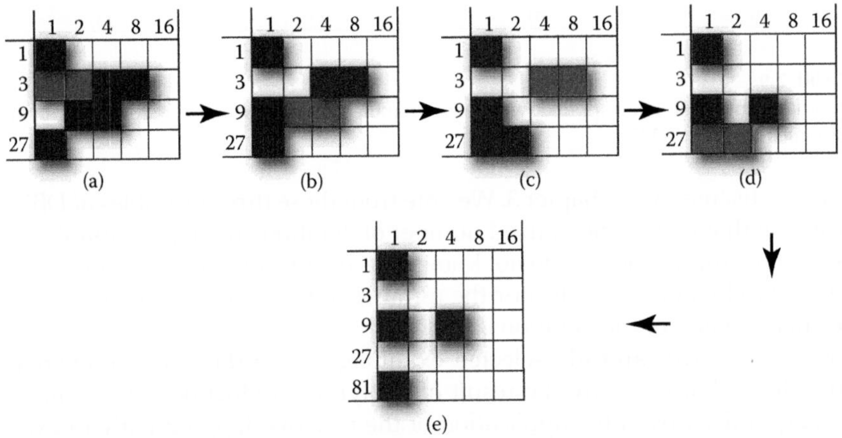

FIGURE 2.11
Applying only the row reduction rule.

call the shift to the right of conflicting digits the *overlay reduction* rule, as given in Equation (2.12).

$$2^i 3^j + 2^i 3^j = 2^{i+1} 3^j \tag{2.12}$$

It would seem reasonable that other, more complex rules can be found to aid in reduction of DBNS representations. In the following subsection we look at the generalized reduction problem.

2.5.3 Generalized Reduction [5]

We can generalize the reduction problem using the purely exponential Diophantine equation (2.13), where $l < k$.

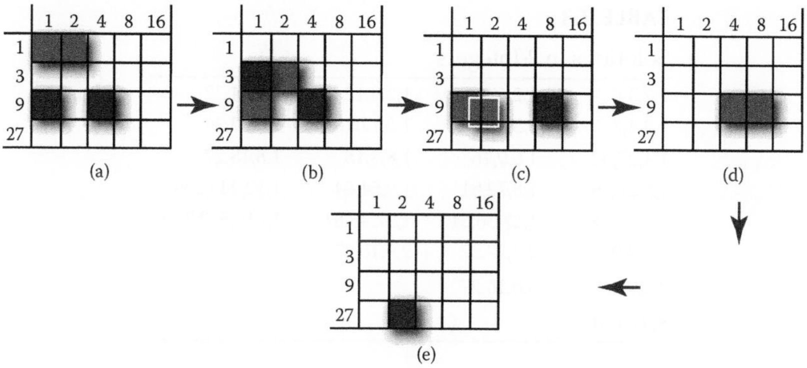

FIGURE 2.12
Applying the row reduction and overlay rules to Figure 2.5 from $x = 41$.

$$\begin{aligned} &2^{l_1}3^{j_1} + 2^{l_2}3^{j_2} + \cdots + 2^{l_k}3^{j_k} \\ &= 2^{m_1}3^{n_1} + 2^{m_2}3^{n_2} + \cdots + 2^{m_k}3^{n_k} \end{aligned} \tag{2.13}$$

Taking the example of $k = 2$, $l = 1$, we have the following theorem [27]:

Theorem 2.2: The Diophantine equation $x + y = z$, where GCD $(x, y, z) = 1$ and (x, y, z) are 6-integers (that is, x, y, and z have the form $2^{x_1}3^{x_2}5^{x_3}7^{x_4}11^{x_5}13^{x_6}$, with $x_i \geq 0$, i = 1, 2, 3, 4, 5, 6), has exactly 545 solutions.

Proof: See [27]. ■

In our case $x_3 = x_4 = x_5 = x_6 = 0$ and the only solutions of $x + y = z$ are {1, 2, 3}, {1, 3, 4}, and {1, 8, 9}. Therefore these three cases represent the only ones where we can replace two active cells with one. We recognize the first case as the row reduction rule (Figure 2.7) and the second case as the column reduction rule (Figure 2.8).

For $k = 3$, $l = 1$ we have a table of useful 3:1 reduction rules from [5], as shown in Table 2.3. Although all of the rules from Table 2.3 can be employed where valid for a particular representation, the complexity of a computational unit that has access to all of the rules will be considerable. Therefore, limiting the number of rules that can be accessed will be prudent for limiting the complexity of an arithmetic unit, even in the trade-off between complexity and speed. Probably the most useful are the rules with closely spaced input digits. The first two rules on the first row of the table satisfy this criterion, as shown in Figures 2.13 and 2.14. As an example of a rule with more widely spaced input digits, Figure 2.15 shows the reduction rule {1, 8, 9, 18} from Table 2.3.

The reduction rules are not only used to increase the sparsity of the representation, but also to guarantee that arithmetic operations can be carried out in a minimum number of iterations. This is discussed in detail in Chapter 3.

TABLE 2.3
Solutions in 2-Integers

1,2,3,6	1,2,6,9	1,2,9,12	1,2,24,27
1,3,4,8	1,3,8,12	1,3,12,16	1,3,32,36
1,4,27,32	1,6,9,16	1,8,9,18	1,8,18,27
1,8,27,36	1,8,72,81	1,9,54,64	1,12,243,256
1,16,64,81	1,27,36,64	1,32,48,81	1,216,512,729
2,3,4,9	2,3,27,32	2,9,16,27	
3,4,9,16	3,8,16,27		
8,9,64,81			

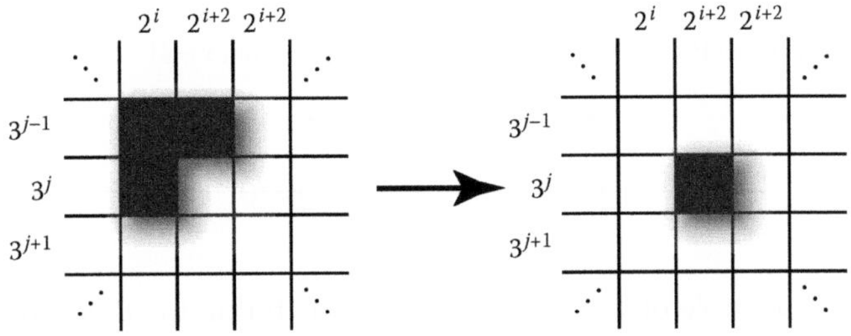

FIGURE 2.13
{1, 2, 3, 6} reduction rule from Table 2.3.

2.6 A Two-Dimensional Index Calculus

For the final part of this introduction to the DBNS, we discuss the representation of a DBNS number via the exponents of the two bases, rather than the position of the digits in a two-dimensional table. We have already used logarithms in Chapter 1, where the slide rule was described as an analog logarithmic processor, and where finite field logarithms—introduced as an index calculus—were used in a number theoretic transform. Arithmetic operations in a DBNS index calculus are carried out using the exponents directly, rather than the techniques discussed in Chapter 3, where direct manipulations of the digits in the tabular representation are used. The equivalent to this is the binary logarithmic number system (LNS), where a number, x, is represented by its logarithm $\log_2 x$, where the arithmetic operations are carried out using the base 2 logarithm representation, rather than x itself.

2.6.1 Logarithmic Number Systems

The logarithmic number system has often been discussed as an alternative to the binary representation. In the LNS, multiplication and division are easy operations, whereas addition and subtraction are difficult operations,

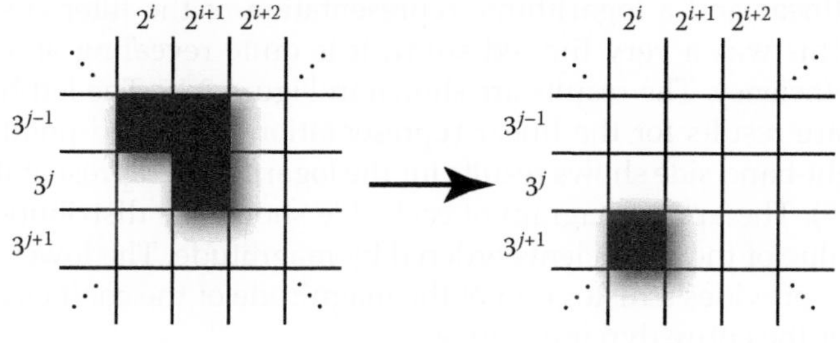

FIGURE 2.14
{1, 2, 6, 9} reduction rule from Table 2.3.

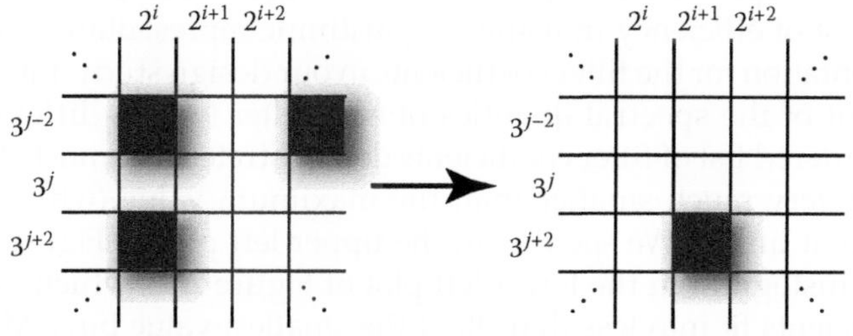

FIGURE 2.15
{1, 8, 6, 18} reduction rule from Table 2.3.

traditionally implemented by making use of large ROM arrays, or other techniques [29,30]. It has been recognized that LNS architectures are perfectly suited for low-power, low-precision DSP problems.

Inner products computed in DSP algorithms are often between a predetermined set of coefficients (e.g., finite impulse response (FIR) filter coefficients or discrete transform basis functions) and integer data. For fixed-point binary implementations, the uniform quantization properties are perfectly matched to the mapping of most input data that use a reasonably large fraction of the representation dynamic range. However, there are cases, such as the mapping of input data for nonlinear hearing instrument processing, where a nonuniform quantization mapping has advantages. For representing predetermined coefficients, a number system that provides a nonuniform quantization mapping strategy will often yield a more efficient use of the representation dynamic range [31].

2.6.2 A Filter Design Study

We can see the effect of changing the number system in the following study of the representation of the coefficients from several different FIR filter designs. In this study we designed eight filters (two of each: lowpass, highpass, bandpass, bandnotch) using Maple-15 (Maplesoft™), and examined the efficiency of both a linear and a logarithmic representation of the filter coefficients. Although this was a very limited study, it is quite revealing about representation efficiency. The results are shown in Figure 2.16. The left-hand side diagrams are results for the linear representation (e.g., fixed-point binary), and the right-hand side shows results for the logarithmic representation (e.g., binary LNS). The upper diagram of each side shows the distribution of the absolute value of the coefficients ordered by magnitude. The lower diagram of each side provides a histogram of the magnitude of the coefficients using 50 bins over the entire dynamic range.

The optimum representation should produce uniformly distributed histograms; however, finding such a representation that would also be efficient at implementing arithmetic with the coefficients is very doubtful. Still, putting the ideal representation problem aside, Figure 2.16 demonstrates the improvement of efficiency in using a logarithmic representation over a linear representation for the filter coefficients in our design study. Interestingly, independent of the spectral qualities of each filter (which differ considerably), the ordered list of filter coefficients demonstrates that most of the coefficients are very much smaller than the maximum value (which we have normalized at unity). We see this in the upper left plot of Figure 2.16, and also in the histogram in the lower left plot of Figure 2.16, which shows that most coefficients fit into less than 10 of the smallest-value bins. Most of the dynamic range of our linear representation is therefore not used to represent the coefficients. The logarithmic representation, on the other hand, is very good at representing quite small values with a large dynamic range, and we can see the effect of this in the plots on the right-hand side of Figure 2.16. The lower right-hand side histogram shows that the coefficient representation spreads over a large portion of the 50 bins, which will provide a more efficient representation of the coefficient sequence than the linear system. Of course, the data that are being filtered are often more efficiently implemented with a linear representation (reflecting on an input signal that is uniformly distributed over the dynamic range). However, there are still compromises that can be made between the representation of the input data and the coefficient sequence in order to improve the representation efficiency of the overall filtering operation.

In order to be able to improve the representation efficiency, we need several degrees of freedom with which to manipulate the representation. For logarithmic type systems (which includes floating point) the LNS and floating point have limited degrees of freedom. In fact, with floating point, the commercial requirement for standardized precision between the mantissa and exponent ranges removes these 2 degrees of freedom from being used to improve the efficiency of the representation for specific applications, such as FIR filters. The LNS, in its basic form, only has the word length and mapping multipliers as parameters to manipulate. In this chapter we will improve on these limits of the degrees of freedom by generalizing the LNS number system and presenting several results

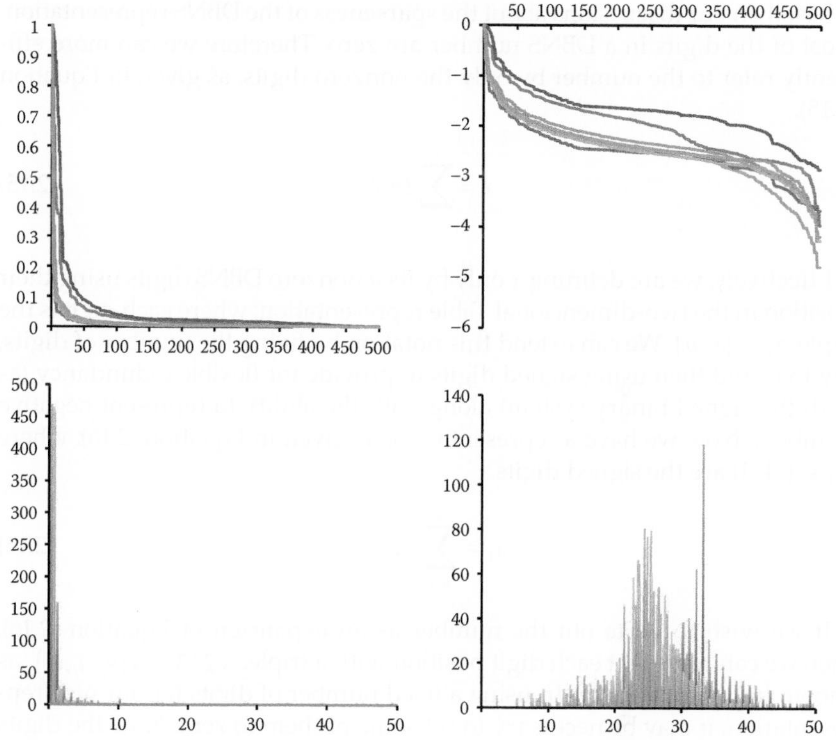

FIGURE 2.16 (See color insert)
Coefficient statistics for eight different FIR filter designs. Number representations: linear (left) and logarithmic (right). Upper plots show the ordered coefficient absolute values, lower plots show 50-bin histograms of the dynamic range usage.

that demonstrate the efficiencies in using this representation over the classical LNS for typical DSP computations. We will also detail the hardware used to perform mapping from binary to the MDLNS along with implementation examples, which demonstrate the efficiencies of using the MDLNS.

2.6.3 A Double-Base Index Calculus

Recall our earlier introduction of the binary-ternary double-base number system, where a number is represented with a series of digits, over two orthogonal dimensions (i, j), having a weight of 2^i3^j, as given by Equation (2.16), where $\{i, j\} > 0$ are integers and $d_{ij} \in \{0, 1\}$ are the digits.

$$x = \sum d_{i,j} 2^i 3^j \tag{2.14}$$

Based on our discussions about the sparseness of the DBNS representation, most of the digits in a DBNS number are zero. Therefore we can more efficiently refer to the number by only the nonzero digits, as given in Equation (2.15).

$$x = \sum_{i=1}^{n} 2^{a_i} 3^{b_i} \tag{2.15}$$

Effectively, we are defining x only by its n nonzero DBNS digits using their position in the two-dimensional Table representation, where each digit is the tuple: $x_i \rightarrow \{a_i, b_j\}$. We can extend this notation by fixing the number of digits, say to n, and then using signed digits to provide for flexible redundancy (as with the signed binary system) along with the ability to represent negative numbers. Now we have a representation, as given in Equation (2.16), where $\{s_i\} \in (-1, 1)$ are the signed digits.

$$x_i = \sum_{i=1}^{n} s_i 2^{a_i} 3^{b_i} \tag{2.16}$$

If we wish to write out the number as an expansion of Equation (2.16), then we can represent each digit position with a triple, $s_i 2^{a_i} 3^{b_i} \rightarrow (s_i, a_i, b_i)$, as shown in Equation (2.17). In using a fixed number of digits for our new representation, it may be necessary to set some of them to zero. Now the digits $\{s_i\} \in (-1, 0, 1)$.

$$x = \sum_{i=1}^{n} (s_i, a_i, b_i) \tag{2.17}$$

We now make two observations:

1. We can expand the tabular representation by allowing the $\{a_i, b_i\}$ to be signed integers.
2. We can perform arithmetic by manipulating the Table indices $\{a_i, b_i\}$ rather than using the two-dimensionally connected logic from Chapter 4.

Observation 1 now allows rational expressions with integers, providing a means of representing real numbers rather than just integers. This also provides the opportunity to trade off the dynamic range of the indices $\{a_i\ b_i\}$ with the number of digits used. Observation 2 provides a two-dimensional equivalent of the LNS mapping from binary, allowing the same type of arithmetic complexity: low-complexity (easy) multiplication and division operations, but high-complexity (difficult) addition and subtraction operations.

For the special case where only a single digit is used, the manipulation of indices, as in the above, is referred to as an index calculus double-base number system (IDBNS) [5]. Therefore the IDBNS is based on a single-digit representation of the form in Equation (2.18):

$$y = s2^a3^b \tag{2.18}$$

where $s \in \{-1, 0, 1\}$ and a and b are signed integers. In this case we have the following theorem [5,3]:

Theorem 2.3: For every $\epsilon > 0$ and every nonnegative real number x, there exist a pair of integers a and b such that the following inequality holds:

$$|x - 2^a3^b| < \varepsilon$$

We may therefore approximate, to arbitrary precision, every real number with the triple $\{s, a, b\}$. Therefore, we may look at this representation as a *two-dimensional generalization* of the binary LNS. The important advantage of this generalization in multiplication is that the binary and ternary indices are operated on independently from each other, with an attendant reduction in complexity of the implementation hardware. As an example, a VLSI architecture for inner product computation with the IDBNS, proposed in [5,8,28], has an area complexity dependent entirely on the dynamic range of the ternary exponents. Providing the range of the ternary exponent is smaller than the LNS dynamic range for equivalent precision, we have the potential for a large reduction in the IDBNS hardware compared to that required by the LNS. We can capitalize on this potential improvement by placing design constraints on the ternary exponent size. For example, if we want to represent digital filter coefficients in the IDBNS, then we can design the coefficients in such a way that the ternary exponent is minimized: an integer programming task [8,28]. Although this approach is sound, and can produce modest improvements, we can do better. In fact, the complexity of the architecture does not depend on the particular choice of the second base. Therefore, one may attempt to find a base, x, or set of bases such that the filter coefficients can be very well approximated by the form $s2^ax^b$, while keeping b (the nonbinary exponent) as small as possible.

This leads us to the multidimensional logarithmic number system (MDLNS), which is discussed in detail in Chapter 5, with applications in Chapters 6–9.

2.7 Summary

In this chapter, we have introduced the double-base number system, the number representation that forms the basis of the remainder of this book. The origins of the DBNS came from an observation about reducing computation time for specific problems in matrix computation associated with graph theory, and modular exponentiation, which is a computationally complex problem in cryptography. The use of the DBNS, and its extensions, in the implementation of architectures for digital signal processing applications, arose out of work conducted in the authors' research groups.

The DBNS is redundant, in fact much more so than signed digit binary, but the DBNS has similar advantages in computation to signed digit binary if the redundancy can lead to sparseness. Canonic forms for DBNS representations are shown to be very sparse, and even though it is a computationally hard problem to find canonic representations, a simple greedy algorithm, with an expansion of sublinear length, is demonstrated to produce near-canonic representations, which can also be very sparse. A simple two-dimensional table, in which the elements of the table are 0 or 1, can be used to show DBNS representations, and using this tabular structure, basic rules are developed to allow a symbolic substitution approach to reducing noncanonic representations to near-canonic ones.

The table approach also leads to the concept of a representation that only uses the table elements that have the value 1. These elements are represented by the indices of their position in the table, leading to an index calculus arithmetic that can be interpreted as a multidigit logarithmic arithmetic over two bases. Also, an illustration is provided to show that logarithmic representations can be very efficient for certain applications—in our example we use the impulse response characteristics of digital filters.

References

1. V.S. Dimitrov and T.V. Cooklev, Hybrid Algorithm for the Computation of the Matrix Polynomial, *IEEE Transactions on Circuits and Systems*, 42(7), 377–380, 1995.
2. V.S. Dimitrov and T.V. Cooklev, Two Algorithms for Modular Exponentiation Using Nonstandard Arithmetic, *IEICE Transactions on Fundamentals*, E78-A, 82–87, 1995.
3. V.S. Dimitrov, S. Sadeghi Emamchaie, G.A. Jullien, W.C. Miller, A Near Canonic Double-Based Number System (DBNS) with Applications in Digital Signal Processing, in *Proceedings of SPIE Conference on Advanced Signal Processing Algorithms, Architectures and Implementations IV*, Denver, August 1996, pp. 14–25.
4. Y. Ibrahim, G.A. Jullien, W.C. Miller, V.S. Dimitrov, DBNS Arithmetic Using Cellular Neural Networks, in *IEEE International Symposium on Circuits and Systems*, ISCAS '05, 2005, pp. 3914–3917.
5. V.S. Dimitrov, G.A. Jullien, W.C. Miller, Theory and Applications of the Double-Base Number System, *IEEE Transactions on Computers*, 48(10), 1098–1106, 1999.

6. R. Muscedere, G.A. Jullien, V.S. Dimitrov, and W.C. Miller, On Efficient Techniques for Difficult Operations in One and Two-Digit DBNS Index Calculus, in *Proceedings of 34th Asilomar Conference on Signals, Systems and Computers*, 2000, pp. 870–874.
7. V.S. Dimitrov, L. Imbert, A. Zakaluzny, Multiplication by a Constant Is Sublinear, in *18th IEEE Symposium on Computer Arithmetic*, 2007, pp. 261–268.
8. G.A. Jullien, V.S. Dimitrov, B. Li, W.C. Miller, A Hybrid DBNS Processor for DSP Computations, in *Proceedings of 1999 IEEE Symposium on Circuits and Systems*, 1999, vol. 1, pp. 5–8.
9. R. Muscedere, V.S. Dimitrov, G.A. Jullien, W.C. Miller, A Low-Power Two-Digit MDLNS Filterbank Architecture for a Digital Hearing Aid, *EURASIP Journal on Applied Signal Processing*, 18, 3015-3025, 2005.
10. V.S. Dimitrov, G. Jullien, Loading the Bases—A New Number Representation with Applications, *IEEE Circuits and Systems Magazine*, 3(2), 6–23, 2003.
11. V.S. Dimitrov, G.A. Jullien, W.C. Miller, An Algorithm for Modular Exponentiation, *Information Processing Letters*, 66(3), 155–159, 1998.
12. V. Dimitrov, L. Imbert, P. Mishra, Efficient and Secure Elliptic Curve Point Multiplication Using Double-Base Chains, in *Advances in Cryptology—ASIACRYPT '05*, 2005, pp. 59–78.
13. V.S. Dimitrov, K.U. Järvinen, M.J. Jacobson Jr., W.F. Chan, Z. Huang, FPGA Implementation of Point Multiplication on Koblitz Curves Using Kleinian Integers, in *Cryptographic Hardware and Embedded Systems—CHES '06*, 2006, pp. 445–459.
14. V. S. Dimitrov, L. Imbert, P. K. Mishra, The Double-Base Number System and Its Application to Elliptic Curve Cryptography, *Mathematics of Computation*, 77(262), 1075–1104, 2007.
15. V.S. Dimitrov, K.U. Järvinen, M.J. Jacobson, W.F. Chan, Z. Huang, Provably Sublinear Point Multiplication on Koblitz Curves and Its Hardware Implementation, *IEEE Transactions on Computers*, 57(11), 1469–1481, 2008.
16. A. Borodin, P. Towari, On the Decidability of Sparse Univariate Polynomial Interpolation, *Computational Complexity*, 1, 67–90, 1991.
17. P. Montgomery, A Survey of Modern Integer Factorization Algorithms, *CWI Quarterly*, 7(4), 337–366, 1994.
18. M.D. Ercegovac, T. Lang, J.G. Nash, and L.P. Chow, An Area-Time Efficient Binary Divider, in *IEEE International Conference on Computer Design*, Rye Brook, NY, October 1987, pp. 645–648.
19. P. Kornerup, Computer Arithmetic: Exploiting Redundancy in Number Representations, presented at ASAP95, Strasbourg.
20. A. Avizienis, Signed-Digit Number Representation for Fast Parallel Arithmetic, *IRE Transactions on Electronic Computers*, 10, 389–400, 1961.
21. H. Garner, Number Systems and Arithmetic, *Advances in Computers*, 6, 131–194, 1965.
22. G.W. Reitwiesner, Binary Arithmetic, *Advances in Computers*, 1, 231–308, 1960.
23. Booth, A.D., A Signed Binary Multiplication Technique, *Quarterly Journal of Mechanics and Applied Mathematics*, 4(2), 236–240, 1951.
24. M.D. Ercegovac, T. Lang, *Digital Arithmetic*, Morgan Kaufmann (Elsevier Science), San Francisco, 2004.

25. V. Dimitrov, E. Howe, Lower Bounds on the Length of Double-Base Representations, *Proceedings of the American Mathematical Society*, 10, 3405–3411, 2011.
26. R. Tijdeman, On the Maximal Distance between Integers Composed of Small Primes, *Compositio Mathematica*, 28, 159–162, 1974.
27. B.M.M. de-Weger, Algorithms for Diophantine Equations, *CWI Tracts–Amsterdam*, 65, 1989.
28. J. Eskritt, Inner Product Computational Architectures Using the Double Base Number System, MASc thesis, Electrical and Computer Engineering, University of Windsor, 2001.
29. D.M. Lewis, Interleaved Memory Function Interpolators with Application to an Accurate LNS Arithmetic Unit, *IEEE Transactions on Computers*, 43(8), 974–982, 1994.
30. J.N. Coleman, E.I. Chester, C.I. Softley, J. Kadlec, Arithmetic on the European Logarithmic Microprocessor, *IEEE Transactions on Computers*, 49(7), 702–715, 2000.
31. H. Li, G.A. Jullien, V.S. Dimitrov, M. Ahmadi, W.C. Miller, 2002, A 2-Digit Multidimensional Logarithmic Number System Filterbank for a Digital Hearing Aid Architecture, in *Proceedings of 2002 International Symposium on Circuits and Systems*, vol. II, pp. 760–763.
32. V. S. Dimitrov, K.U Järvinen, J. Adikari, Area-Efficient Multipliers Based on Multiple-Radix Representations, *IEEE Transactions on Computers*, 60(2), 189–201, 2011.
33. R. Muscedere, V.S. Dimitrov, G.A. Jullien, Multidimensional Logarithmic Number System, in *The VLSI Handbook*, 2nd ed., W.-K. Chen (ed.), CRC Press, Boca Raton, FL, 2006, pp. 84-1–84-35.

3

Implementing DBNS Arithmetic

3.1 Introduction

In Chapter 2, we introduced the concept of representing numbers over two bases, rather than the ubiquitous single-base (single-radix) positional number representation. Although the representation was originally introduced for reducing the complexity of matrix computations for applications such as computer graphics [1], and large integer computations over finite fields, for cryptography applications [2], our research group initially adopted it for digital signal processing (DSP) applications. This new representation fits in well with a targeted interest in alternative techniques for building DSP processors on silicon.

Although our main interest in the double-base number system (DBNS) for DSP applications turned out to be in using index calculus (which gave rise to the multidimensional logarithmic number system (MDLNS)—Chapters 4–9), the concept of using symbolic substitution to compute with the two-dimensional representation, introduced by one of the authors (Dimitrov), and a long-term interest in image analysis and filtering within our group, provided a connection to a recent development of cellular neural network (CNN) processors [3,4]. In particular, the use of CNNs to perform image morphology processing [5], such as pixel dilation and erosion, appeared to be perfectly suited to the problem of reduction of nonzero digits in two-dimensional DBNS maps. In our research group we have explored such image analysis paradigms in a variety of arithmetic applications, including standard and signed digit binary, as well as DBNS. Our work has included integrated circuit designs for the CNN processors, where the networks are built as self-timed locally connected nonlinear (saturating) analog circuits (cells in the CNN array). Other groups have also considered using the DBNS in CNN implementations of arithmetic computation [6]. In a computational paradigm, a CNN array can be put in the same general class as the well-researched asynchronous (self-timed) logic [7].

3.1.1 A Low-Noise Motivation

If one looks at standard digital logic gates as nonlinear saturating high-gain amplifiers (certainly complementary metal oxide semiconductor (CMOS) falls into this classification), then it seems reasonable to expect that CNNs can also be used to implement logic. There are, however, some major differences:

1. CNN cells use feedback from a neighborhood of cells, including the cell itself, in order to implement the processor function(s) of the array. In contrast, the use of feedback in combinatorial logic circuits requires clocked registers to prevent races.
2. Networks of synchronous logic require global clocks that require carefully designed networks in order to minimize timing skews in the clocks across the network [8]. Self-timed logic is asynchronous and does not have such a timing problem.
3. Large synchronous logic arrays pull considerable current out of the power supply to the array when the clock changes state, causing substantial *Ldi/dt* voltage changes across parasitic inductance in series with the power supply. This is not a problem with self-timed arrays.

The latter observation provides a motivation for pursuing a CNN solution to arithmetic processing. Lossy integration is built into the CNN cell, similar to the use of integrators in analog computers (see Chapter 1), to implement the state equations that define the cell operation. The integrators in each cell of an array can be connected to a global control that changes the time constant associated with the integrator in each cell. This has the effect of "smoothing out" the current and voltage transitions as the array responds to its initial state and its inputs and feedback from the cell neighborhood, including itself. In contrast, logic gates are designed to switch as quickly as possible, in order to reduce combinatorial circuit delay. The transitions of current flow into CMOS logic gates are therefore quite abrupt.

The rapid changes in current flow that induce large voltage changes can cause substrate currents to flow, with the potential to change logic levels in registers, and thus create errors that propagate through the circuitry [9]. If we reduce clock frequency, the noise variance drops, but the instantaneous substrate current injection is unaltered, it just happens at a lower repetition rate. This property provides an extra bonus in exploring the utility of a different paradigm (analog CNNs) to implement arithmetic logic. The reason for this is that a single control on the time constant of every cell in the CNN array provides a global control over the instantaneous noise that the array generates as it performs its computations.

3.2 Arithmetic Operations in the DBNS

In Chapter 2 our introduction to the double-base number system (DBNS) included a formal introduction, an algorithm to convert normally represented numbers into the DBNS (the greedy algorithm), a tabular method to display the representation, and some basic reduction rules using the tabular representation. In this and the following sections we will discuss the implementation of arithmetic (specifically addition and multiplication) using the two-dimensional tabular representation, and the reduction rules developed in Chapter 2, to explain the various techniques we have developed.

In this section we introduce basic methods for addition and multiplication using the tabular approach with substitution rules, such as those used above for digit reduction.

3.2.1 DBNR Addition

As a starting point, assume that we are adding two numbers, $x + y = z$, where both have a canonic representation. Let $I_x(i, j)$, $I_y(i, j)$, and $I_z(i, j)$ be the DBNS maps of the integers x, y, and the output sum, z, respectively. The addition is carried out by simply concatenating the DBNS maps of x and y, and correcting for any overlays of the same digit in both DBNS maps. We note that if x and y both contain the digit 2^i3^j, digit $2^{i+1}3^j$ must be zero; otherwise we could apply the row reduction rule (from Chapter 2) to remove the digit. This is clearly not possible if the representations are canonic as assumed. This means that we can directly apply the *overlay reduction rule* from Chapter 2, Equation (2.12), with the reduced single digit placed in the zero digit location. Both the concatenation and overlay operations are handled using Equations (3.1) and (3.2).

$$I_z(i+1, j) = I_x(i, j) \text{ AND } I_y(i, j) \tag{3.1}$$

$$I_z(i, j) = I_x(i, j) \text{ XOR } I_y(i, j) \tag{3.2}$$

We can relax our assumption of canonic DBNS representations for x and y by exhaustively applying the row reduction rule to both representations before performing addition. We will refer to a map where no further row reduction can take place as an addition-ready DBNS representation (ARDBNR). It is clear that the row reduction rule is the only rule we need to apply in order to perform addition, though other rules, in addition to row reduction, could be used to speed up the reduction process.

We demonstrate these concepts with two different reductions of the addition 1091 + 1501 = 2592. Since 2592 = 81 × 32, the canonic form will be a single digit at the sixth column and fifth row of the table structure we have been using. Figure 3.1 shows the iterations used in row reduction and overlay reduction only. We have selected DBNS representations for 1,091 and 1,501 so that we obtain a concatenated representation that starts the carry propagation process and allows it to continue through the row reduction rule. As with Figure 2.9 in Chapter 2, we show the two procedures (row then overlay

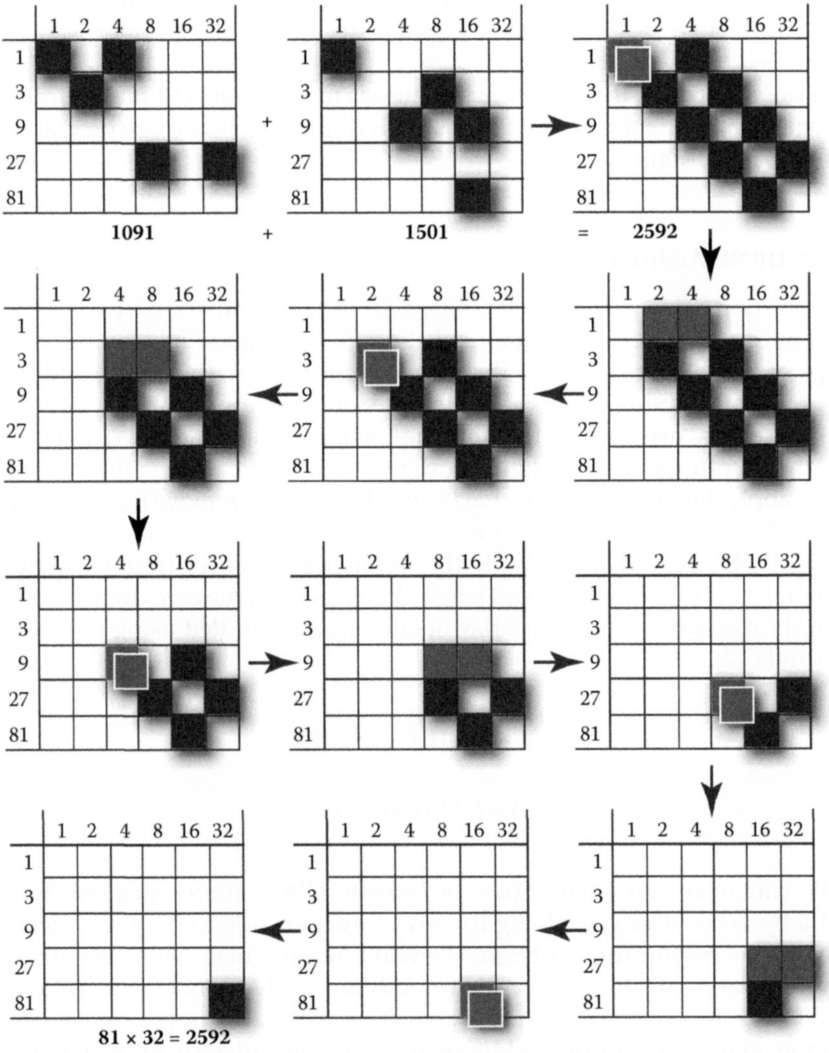

FIGURE 3.1
Demonstrating addition and carry propagation through the row reduction rule.

reduction) in adjacent tables even though we count the two rule applications as a single carry propagation mechanism. The carry propagation is started by the application of Equations (3.1) and (3.2), essentially concatenation followed by overlay reduction. In contrast to this example, we show the same addition in Figure 3.2, but using the column reduction rule from Chapter 2. In this case, the carry propagation terminates after the initial concatenation and overlay reductions. The column reduction only continues for another two steps before there is no other simple reduction rule step possible, so the representation is ARDBNR. In both Figures 3.1 and 3.2, the reduction rules were applied until the representation was ARDBNR, and so we are tempted to conclude that the latter example is more efficient than the former. This is somewhat misleading, however, since the canonic result from Figure 3.1 is maximally sparse and clearly a better choice for further computations (i.e., will probably result in much fewer reduction rule applications in the following computation) than the result in Figure 3.2.

As an illustration, taking Figure 3.1 as a worst case for application of row reduction, we can compare the effective carry propagation of the number of columns in the DBNS table with that for a K-bit binary number with a worst-case propagation of K-bit positions. For a DBNS table with B rows and A columns, we will assume, as with the K-bit number, that all digit positions are filled. We therefore have a filled table with the value given by Equation (3.3).

$$\sum_{i=0}^{B-1} 3^i \sum_{j=0}^{A-1} 2^j = \frac{1}{2}(2^A - 1)(3^B - 1) \tag{3.3}$$

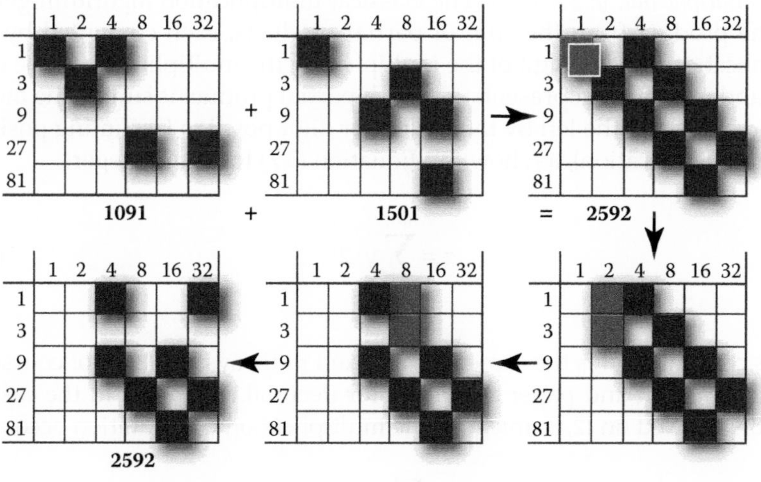

FIGURE 3.2
Early carry propagation termination through the column eduction rule.

Taking the maximum DBNS number as larger than or equal to the maximum binary number, then:

$$2^K - 1 \le \frac{1}{2}(2^A - 1)(3^B - 1) \tag{3.4}$$

For large numbers we can make the approximation:

$$2^K \le 2^{A-1}3^B \tag{3.5}$$

Taking logs and manipulating, we have:

$$K + 1 - B\log_2 3 \le A \tag{3.6}$$

This inequality gives a worst-case "carry" propagation for a table with a flexible aspect ratio A/B. Assuming that the table is square (i.e., $A = B$), then $A = \lceil (K+1)/(1+\log_2 3) \rceil$. For a 12-bit binary number, the worst-case Row Reduction "carry" propagation is $A = \lceil 13/(1+\log_2 3) \rceil = 6$. In Figure 3.1 (which is not quite square) we have five carries. In this case, we see that the number of worst-case carries, which provides a canonic result—only one digit, is reduced by more than 50% compared to the worst-case carries for an equivalently valued binary number system.

3.2.2 DBNS Multiplication

We devote the next chapter to an in-depth discussion about multiplication associated with the DBNS. In this section we provide a few simple examples that extend the previous discussion on addition.

Basic multiplication algorithms produce a product, z, from a *multiplier*, x, and a *multiplicand*, y: $z = x \times y$. The classical multiplication algorithm generates the product from the sum of partial products, each computed by the multiplication of each digit of the multiplier by the multiplicand. In the case of a standard binary representation, each partial product is simply the multiplicand shifted to the left by the multiplier digit position (assuming positive integer multiplication), as shown in Equation (3.7) for B-bit integers.

$$z = \sum_{i=0}^{B-1} y \cdot 2^i \tag{3.7}$$

This computation is the same for any fixed radix system (and, of course, is the classical pen and paper algorithm for decimal numbers). In the case of the DBNS, based on {2, 3}-integers, the multiplication computation becomes:

$$z = \sum_{i,j \in D} y \cdot 2^i 3^j \tag{3.8}$$

where D is the domain of nonzero digits in the DBNS representation of x. In computing each partial product we require a two-dimensional shift in the table representation of y. The summation will involve a repetition of the concatenation/reduction technique discussed in the previous section.

An example of the multiplication of $17 \times 20 = 340$ using a summation of partial products is shown in Figure 3.3. The ARDBNR of the multiplier, 17, the multiplicand, 20, and the product, 340, are shown on the top row (these representations are not canonic). The second row shows the three partial products (two-dimensional shifted versions of 20) and the bottom two rows show the concatenation, overlay, and reduction steps in summing the three

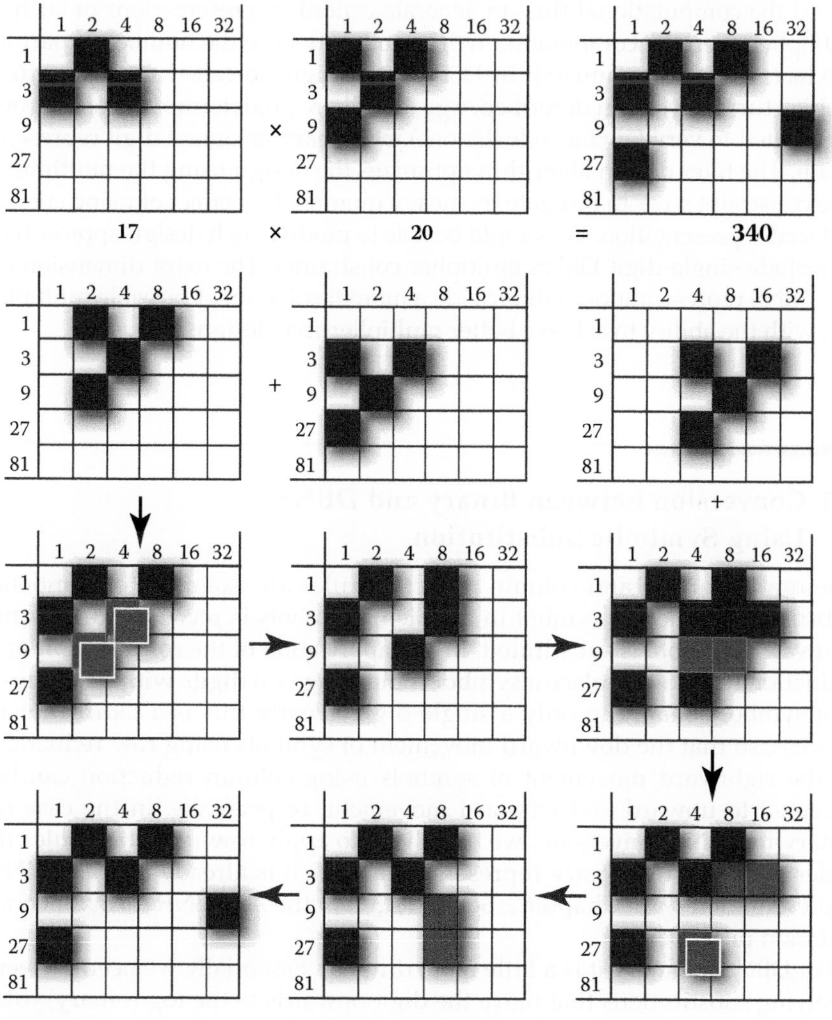

FIGURE 3.3
Multiplication of $17 \times 20 = 340$ using summation of partial products.

partial products. Note, on the third row, that the left-most 2 partial products are overlaid first followed by the overlay of the third partial product. We have used row reduction except for the final stage, where column reduction was used to further reduce the result. Note that we still have an ARDBNR result even without that final column reduction step.

3.2.3 Constant Multipliers

Clearly, if we can guarantee that the inputs to the multiplier are canonic, then the computation will be reduced from that shown. If we are using precomputed multipliers (see a discussion of this in Chapter 1), and we are able to afford the computational time to generate canonic representations of such a multiplier, then the computation will be reduced since the number of partial products will be minimized. In DSP filter design, so-called multiplier-free architectures have been described, e.g., [11], where predetermined coefficients are limited to very few bits (ideally one) in a binary or signed digit representation. The filter design algorithm optimizes the design using the multiplier-free constraint so as to mitigate the heavy quantization effect of using such a reduced representation. We should be able to modify such design approaches to include single-digit DBNS multiplier constraints. The extra dimension of the DBNS representation will provide a more flexible set of possible multipliers, with the ability to achieve better multiplier-free designs.

3.3 Conversion between Binary and DBNS Using Symbolic Substitution

The row reduction and column reduction rules are examples of symbolic substitution, where a symbol in a table of symbols is recognized and the equivalent symbol is substituted. In our application of the row and column reduction rules, we replace a symbol containing two digits with an equivalent symbol containing only a single digit. Clearly this is a bidirectional process, so that the downward movement of symbols using row reduction or the rightward movement of symbols using column reduction can be reversed to upward and leftward movement, respectively. In the case of binary-to-DBNS conversion, we only have to apply row reduction rules in order to obtain the binary representation (which is already a valid DBNS representation—see Chapter 2, Section 2.3.1) in the ARDBNR form for computation purposes.

For DBNS to binary it is a little more difficult. Essentially we need to keep applying substitutions that move the digit upwards to the top (binary) row. Reversing row reduction will produce two digits from one with a movement upwards. Since the input is only a single digit, this rule can be applied until we have a true binary form along the top row. Reversing column reduction will actually produce a digit below the input digit in the table, which is not useful. We can also look at other substitution rules from Table 2.3, Chapter 2, that provide a strong upward movement. An example of this is {1, 16, 64, 81}, which provides an upward movement of four rows for all three output digits. If we compare the reverse of Figure 2.7, Chapter 2 (row reduction), with that of Figure 3.4, we can visualize a trade-off between the upward movement and the connectivity neighborhood of the required substitution, so that although reverse row reduction only produces a single-row upward movement, it has a minimally connected neighborhood (one column and one row), whereas the reverse of {1, 16, 64, 81} requires a six-column and four-row connected neighborhood. The next section on using CNN arrays for the implementation will reveal the neighborhood connectivity problems. As the rows move upwards to the top (binary) row, there will clearly be a "collision" of carries as the single DBNS digits are converted into multiple digits. One approach to handling such a collision is to employ a multidigit carry save register that will be evaluated as a multiple vector carry save binary addition, such as a column compression scheme [10].

As an example, let us consider converting the output of Figure 3.3 to binary. Here we will use the reverse row reduction rule (note that the reverse {1, 16, 64, 81} rule cannot be applied in this example). We will use three carry save rows above the DBNS table and use *carry generation* rules to manipulate the carry save rows; the rules are shown as symbolic substitutions in Figure 3.5. Note that the vertical axis is labeled as 1 (i.e., 3^0) to match the vertical row designation in the original DBNS table. Figure 3.6 shows the complete substitution procedure using only reverse row reduction and the carry generation rules. As in the other examples, we show the selected digits for reduction or carry generation rule application in the table preceding the application of the rule(s). We also identify operations that can be carried out in parallel without interfering with each other.

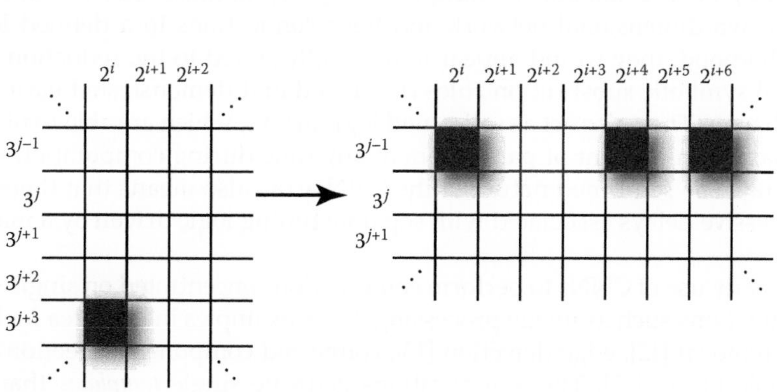

FIGURE 3.4
Reverse reduction rule {1, 16, 64, 81}.

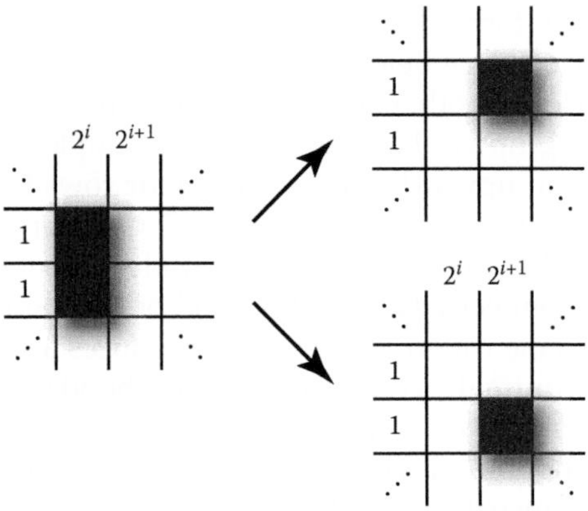

FIGURE 3.5
Carry generation rules.

We have purposely defined the well-known binary carry generation logic as a symbolic substitution in order to provide a homogeneous approach to the implementation. We discuss our unusual arithmetic implementation technique in the following section.

3.4 Analog Implementation Using Cellular Neural Networks

3.4.1 Why CNNs?

Cellular neural networks (CNNs), sometimes referred to as cellular nonlinear networks, were first described (perhaps "invented" might be more appropriate) by Leon Chua and L. Yang in 1988 [3,12]. Because they are defined over a two-dimensional network and have connections to a defined local neighborhood, they would appear to be ideally suited to the reduction and general symbolic substitution rules developed and demonstrated earlier in this chapter. They also act as self-timed logic arrays, which are able to utilize the maximum amount of parallelism at any time during computation with the array. The self-timed nature of the CNN array also means that there are no excessive delays associated with separate timing logic driven by a master clock.

The early use of CNNs to perform computation concentrated on single specific functions such as image processing. Early examples in this area include noise removal [12], edge detection [13], connected component detection [14], and hole filling [15]. These applications generate single *templates* that are used to define the function processed by the network. The next step, that is, creating algorithms that use a programmed combination of such templates,

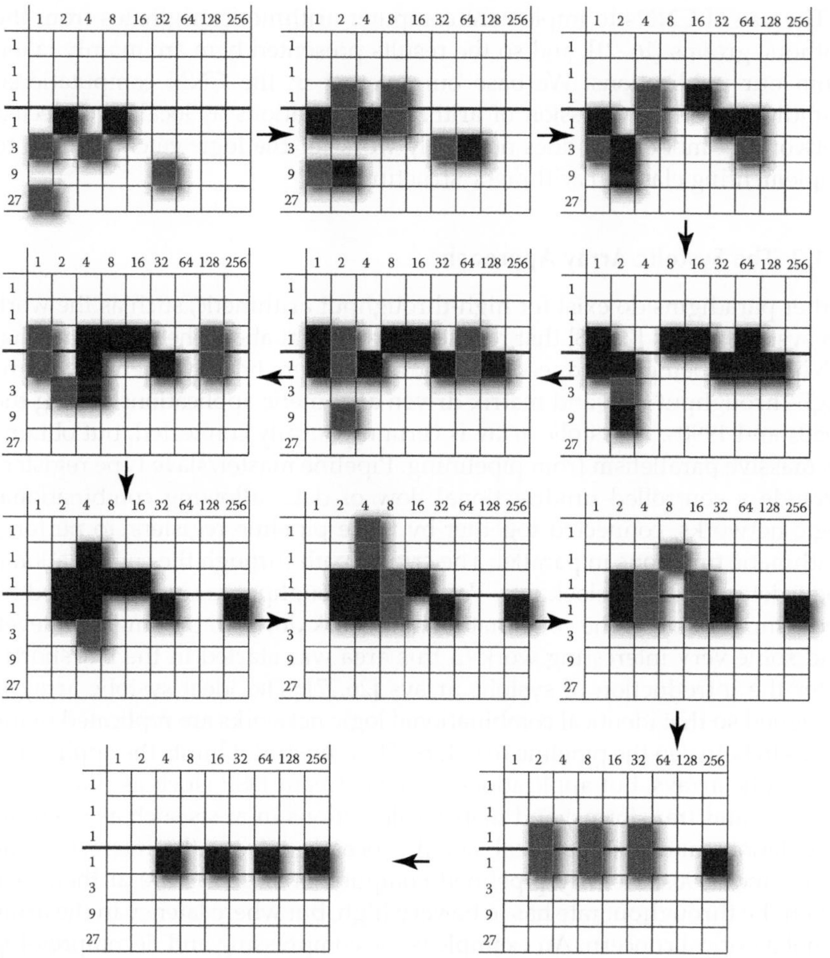

FIGURE 3.6
Converting 340 in DBNS to binary using carry save rows.

was published in 1993 [4], and defined a CNN universal machine in which a set of templates can be controlled externally to produce a desired algorithmic procedure. The interest in CNNs continued apace from that time, and a wealth of examples and research directions can be found in the literature. Our interest is in producing an algorithmic CNN in which the rules we have developed earlier in this chapter are coded into templates and self-timed and controlled to correctly evaluate the templates at any stage in the computational process. This is different from the original universal CNN concept, in that the control of the application of templates is generated from sensing conditions directly within the array rather than having instructions (CNN software) from an external controller (i.e., outside of the array).

The use of CNNs to implement computer arithmetic originates from the authors' groups [16–21], and so the results presented here are mainly taken from our publications. We base our interest in the CNN computational paradigm on the expression of arithmetic operations as locally connected networks, a model that does not really work for the logic gate approach to implementing classical arithmetic structures.

3.4.2 The Systolic Array Approach

Other paradigms do exist for high-throughput arithmetic, such as the work on systolic arrays [22–28] that reached its zenith at about the same time that CNNs were being introduced. Systolic arrays were heavily oriented toward high-throughput DSP and matrix-driven arithmetic applications during the 1980s and 1990s. A systolic array is certainly locally connected, but obtains its massive parallelism from pipelining. Pipeline master/slave type registers provide a controlled unidirectional flow of data, allowing combinational logic networks, connected together by these pipeline registers, to perform arithmetic functions in parallel. The critical path through the combinational logic determines the clock rate. Very high-throughput arrays are formed by reducing the size of the combinational network down to the single-bit level, and some very interesting work in this area was started in the UK shortly after the introduction of systolic arrays [26,27]. The ideal systolic array is designed so that identical combinational logic networks are replicated many times in between the pipeline registers. This constraint limits the application of systolic arrays, but applications in data stream DSP (such as filters, correlators, and transforms) and matrix calculations in areas such as scientific calculations and mathematical signal processing drove the interest in the 1980s and 1990s. Heavily pipelined computational arrays are at their best when the throughput rate has to be very high, but where latency in the array is not a critical concern. An example is the compressing and decompressing of streaming video as heavily used over the Internet. What is important is the rate at which the video data can be processed; the delay before the output appears is not normally a concern.

3.4.3 System Level Issues

Logic gates are highly nonlinear (ideally discontinuous) devices, whereas CNNs can be much smoother nonlinear devices; i.e., their transfer functions are continuously differentiable [29]. This difference brings up VLSI system level phenomena: synchronization and switching noise.

Controlling large VLSI arrays with a master clock has synchronization problems. The early approaches used clock distribution networks [30] in which the master clock is transmitted over a clock tree network that provided nearly identical circuit parameters from the clock to the receiving logic over the entire chip in order to guarantee synchronicity. Differential distribution networks and phase-locked loop buffers represent some of the latest techniques [31]. For the latest multibillion transistor designs, blocks on the chip are connected together as a communication network (network-on-chip [32]), but the basic operation at some subblock levels still relies on the distribution of clocks to synchronous logic networks. Global synchronous clock distribution, along with the issues of achieving synchronicity over a large logic network, affect issues such as high power consumption and dissipation, and switching noise, which can be so large as to connect circuits, designed to be isolated from one another, through the chip substrate and to affect voltage levels in the power distribution network [33].

Removing the synchrony requirement by using self-timed logic has been around for many years, and continues to be a healthy area of research [34]. However, problems related to providing solid design tools to the VLSI logic circuit design community and providing solid testing strategies are some of the issues holding back the widespread adoption of this different paradigm. This is not to say that self-timing cannot be successfully used at the lower levels in VLSI hierarchical design strategies. This is the level at which we would expect to be able to use CNN arrays.

3.4.4 CNN Basics

Although CNNs can theoretically operate over multiple dimensions, for our applications we will restrict the arrays to two dimensions. A CNN consists of an array of connected cells, and the connection radius (neighborhood) is defined as the maximum displacement of a driven cell (a cell receiving an input) from a driver cell (a cell providing an output). In general, each cell will be connected to every other cell in its neighborhood. Initial conditions are placed on the array at the start of a computation, and the array relaxes to a steady-state condition with the results of the computation available on the nodes of the array. For example, if the array is used to manipulate pixels in an image, each pixel value is used as an initial condition for a cell position corresponding to the pixel position in the image. All cells in the CNN operate in parallel, and even though the connection radius may be as small as 1, the array can perform global processing on the entire array given the appropriate algorithm templates.

A CNN array is shown diagrammatically in Figure 3.7 with darkened squares outlining connection radii of 1 and 2. The cells around the border are dummy cells that effectively constrain the network connections to the finite size of the array and ensure proper convergence of the array. We show only a single dummy cell width, suitable for a radius 1 array. The dummy cells output a constant voltage appropriate for ensuring convergence, and they also receive input signals in the same way as do active cells in the array. The dummy cells do not contribute to the final state analysis. For hardware implementations, the convergence radius is kept low (1 or 2) in order to limit the interconnect network. This is possible because the missing global connections

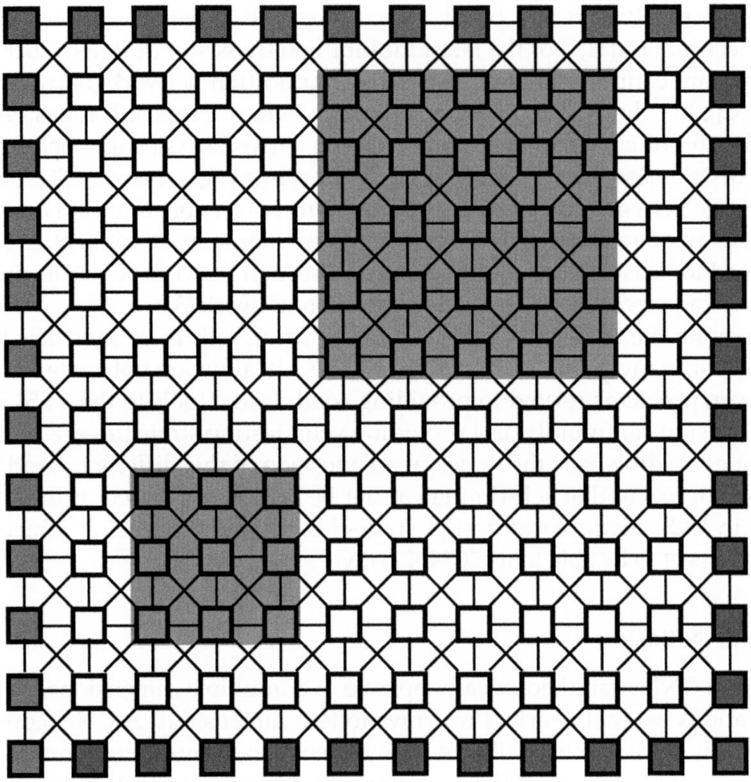

FIGURE 3.7
Cellular neural network array with connection radii of 1 and 2.

are replaced by a time multiplexing of the connections and the time propagation of the information through the network from cell to cell [21].

3.4.5 CNN Operation

We can regard a CNN array as a connected set of cells, that have multiple inputs, a single output, and consist of linear and nonlinear circuit elements. We use a state variable approach to analyzing each cell, where the cell at position i, j in the array has a bounded internal state variable, x_{ij}, external inputs, u_{ij}, and an output y_{ij}. The dynamics of the cell are given by Equations (3.9) and (3.10).

$$C_x \dot{x}_{i,j}(t) = -\frac{1}{R_x} x_{i,j}(t) + \sum_{c \in N_{ij}} A_c \cdot y_c(t) + \sum_{c \in N_{ij}} B_c \cdot u_c(t) + I \tag{3.9}$$

$$y_{i,j}(t) = f\left(x_{i,j}(t)\right) \tag{3.10}$$

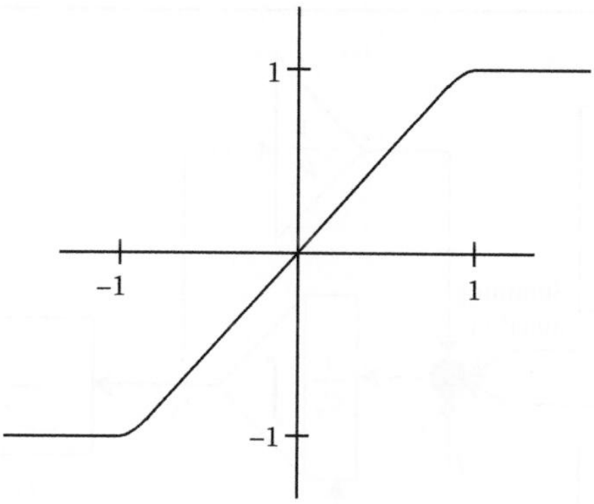

FIGURE 3.8
Nonlinear CNN output function, $f(x_{ij}(t))$.

where I is the bias on the cell (a local value) and N_{ij} is the neighborhood of connected cells. The output function (cell activation function, $f(x_{ij}(t))$ has a nonlinearity, as sketched in Figure 3.8; it is essentially a linear function with a "gain" of unity in the range $(-1 + \varepsilon, 1-\varepsilon)$, and it saturates outside of the range $(-1-\varepsilon, 1 + \varepsilon)$. A_c and B_c are templates for intercell connection weights and input connections, respectively. The A_c template controls the feedback to the cell, and B_c serves as an input control mechanism. Both templates are usually the same for each cell (often referred to as cloning templates). I represents a bias to the network and determines the transient behavior of the network. R_x and C_x form an integration time constant, R_xC_x, that controls the speed of the cell convergence. As discussed in the introduction to this chapter, this time constant can be chosen to change the switching noise of the circuit and provides a trade-off parameter between instantaneous switching noise and computation speed. This trade-off is continuously variable, unlike in logic gate networks, where the instantaneous switching speed of a combinatorial network is fixed. To reiterate the discussion in the introduction, the steady-state noise level of a combinational network can always be lowered by reducing the rate at which inputs are available at the input registers, but the instantaneous noise will be unaltered. If the concern is with the potential for substrate cross talk to change register values, then the instantaneous noise level is a very important parameter.

3.4.6 Block and Circuit Level Descriptions of the CNN Cell

A block level representation of a CNN cell is shown in Figure 3.9. The diagram is reminiscent of the programming techniques for analog computers, where the differential equation is integrated and then implemented with amplifiers and integrators (amplifiers with capacitor feedback). Note the term *summing junction*, which refers to the summing of currents at the input to an operational amplifier with feedback. For very high-input-impedance amplifiers, there will be negligible current flowing into the amplifier, and so,

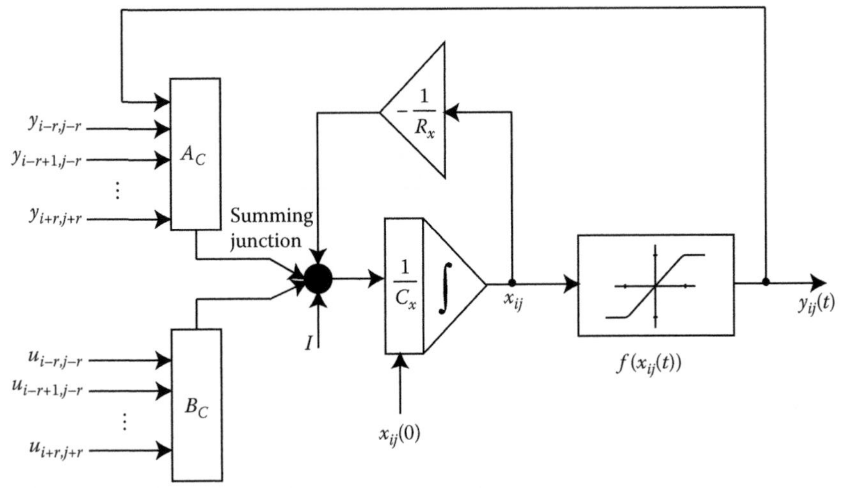

FIGURE 3.9
Block diagram of a CNN cell.

from Kirchoff's current law, whatever current flows into the junction from the inputs must flow out of the junction through the feedback network. With this approach, we are able to build a very accurate inverter. Figure 3.10 shows a typical integration circuit using an amplifier with high-input impedance and high gain; that is, an *operational amplifier* (a term coined in 1947 to refer to high-gain amplifiers). The amplifier amplifies the voltage differential between the two inputs, with polarity as shown. If we make the assumption that the amplifier has infinite gain, then for a finite output voltage, v, the input voltage must be zero. Since the + input is connected to ground, the – input is also at ground potential, even though it is not connected to ground—a virtual ground. We can use this knowledge to both convert voltage on the input resistors to current, $\sum_{j=i}^{n} i_j$, and determine the output voltage. In fact, the output voltage appears directly across the capacitor, C, and so $-\sum_{j=1}^{n} i_j = C\,dv/dt$. Integrating, we obtain the transfer function:

$$v = -\frac{1}{C}\int\left(\sum_{j=1}^{n} i_j\right)dt + v_0 \tag{3.11}$$

where v_0 is the constant of integration (the initial voltage across the capacitor).

We can now see that integrating Equation (3.9) and substituting into Equation (3.10) yields:

$$y_{ij}(t) = f\left(\int\left(-\frac{1}{R_x}x_{ij}(t) + \sum_{c\in N_u} A_c \cdot y_c(t) + \sum_{c\in N_u} B_c \cdot u_c(t) + I\right)dt + x_{ij}(0)\right) \tag{3.12}$$

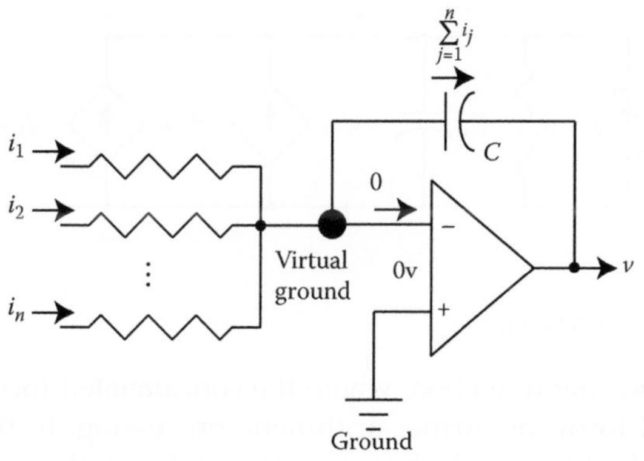

FIGURE 3.10
An integrator circuit.

which is of the same form as the integrator of Equation (3.11), and hence the implementation as shown in Figure 3.9. $x_{ij}(0)$, the constant of integration, is the starting point (an initial charge on the integrator capacitor), that will be used to initialize the CNN array prior to its evaluation through relaxation to an equilibrium value. Rewriting the original differential equation in the integration form eliminates the need to build differentiator circuits, which are unfortunately wonderful noise amplifiers!

Components used in the commercial analog computers of the 1950s to early 1970s were based on high-performance (and very expensive) amplifiers. Such accurate components were necessary in order to provide accurate tracking of solutions to the differential equations being evaluated. We are only interested in the final equilibrium state of the network, when all cells are in one of their two saturated states. The accuracy, in time, of how they relaxed to that state is not important. Based on this fact, the block components in Figure 3.9 can be implemented by the equivalent circuit of Figure 3.11. The integrator function is provided by C_x and R_x; this will act as a "leaky integrator" with losses attributed to R_x. $A_C y_C$ and $B_C u_C$ are linear voltage-controlled current sources that are connected to all cells in the neighborhood, and are driven by the state variables from, and the inputs to, the neighborhood cells, respectively. I_y is a current source controlled by the state voltage, x_{ij}, where the output voltage, y_{ij}, is developed across resistor, R_y; saturation occurs at the power supply voltages. I is an independent bias current source.

3.4.7 DBNS Addition CNN Templates Using Overlay and Row Reduction Rules

In Section 3.2.1 we demonstrated that addition in the DBNS can be carried out by concatenating the DBNS representations of the addends followed by the use of two symbolic substitution rules: overlay, where overlaid digits

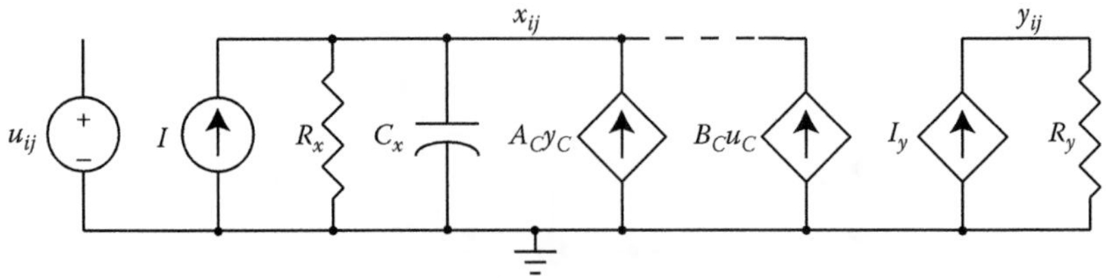

FIGURE 3.11
Equivalent circuit of a CNN cell.

are handled, and row reduction, where the concatenated form is converted to an ARDBNR form for further arithmetic processing. In this section we discuss the method by which the two rules can be implemented in a CNN cell; the input to the addition computation is assumed to be in an ARDBNR form. Details of this procedure can be found in [21], the PhD thesis of Dr. Y. Ibrahim, a former student from our research group at the University of Windsor.

For a CNN implementation, we require to design two templates: one for the overlay rule and the second for the row reduction rule. The operation of the templates is usually handled using an external control unit that is programmed to load different templates in a specific order for certain periods of time [4,35,36]. We had initially investigated a technique [37] that used discrete logic gates attached to each cell to control template operation. In [20,21], a novel self-programmable analog CNN array that implements the two rules was successfully designed and simulated at the circuit level, and we discuss that technique for the remainder of this chapter. In brief, this CNN network switches between the overlay rule and the row reduction rule, based on the output voltages of the involved cells without the need to use external digital control logic. This produces a design that is entirely analog, and which will provide the self-timed, low-noise advantages discussed in Section 3.4.5.

3.4.8 Designing the Templates [21]

In the CNN paradigm, developing a certain function or operation requires the design of templates and their connections to neighboring cells: the output feedback template, Ac, the input control template, Bc, and the bias to each cell, I, so that when applying these templates to the CNN network, the output of the network corresponds to the desired function applied to the input. Implementing logic functions is an established CNN design process [38], and we use this approach to implement the symbolic substitution rules that have been demonstrated in the earlier sections of this chapter. As an example, in Section 3.2.1 the DBNS rules of concatenation and overlay were handled using the Boolean logic equations (3.1) and (3.2).

For the row reduction rule template we start with Equation (2.10) from Chapter 2, repeated below for convenience:

$$2^i3^j + 2^{i+1}3^j = 2^i3^{j+1} \tag{3.13}$$

Examining the graphical representation of the reduction rule in Figure 2.7, Chapter 2, reveals that the cell at position $(i, j+1)$, $I_z(i, j+1)$ should be activated (set to a nonzero value) if and only if the two cells at positions (i, j) and $(i+1, j)$ are active. The operation effectively implements a two-input logical AND function, with the output voltages of the cells $I_z(i, j)$ and $I_z(i+1, j)$ representing the inputs to the AND function and the output voltage of the cell $(i, j+1)$ representing the output of the AND function. Unlike standard logic equations, where there is no explicit timing information (a logic equation is valid even if there is no delay between applying the inputs and measuring the output from a gate), the CNN approach is derived from state equations (3.9) and (3.10), where time is a fundamental variable. To recognize this in the DBNS arithmetic logic equations, we can first introduce a superscript notation to separate prior (superscript 1) and post (superscript 2) values on the cell nodes. Thus the output voltage of cell $(i, j+1)$ can be described using the equivalent logic equation with the digits in the DBNS representation:

$$I_z^2(i,j) = I_z^1(i,j-1) \text{ AND } I_z^1(i+1,j-1) \tag{3.14}$$

Using the notation introduced in Section 3.4.5, we use Equation (3.14) to derive a continuous time feedback template, as shown in Equation (3.15).

$$A_{i,j}(y_c(t)) = \beta \cdot (y_{i,j-1}(t) + y_{i+1,j-1}(t) - 1) \tag{3.15}$$

In understanding this equation, we recognize that all cell outputs are operated on by the nonlinear (saturation) function of Figure 3.8, and so cannot exceed the range (–1, 1). Since a cell is inactive when the output is saturated at –1, and active at + 1, we see, given $\beta = 1$, that the feedback is only active when both inputs are saturated at + 1. If only one of the inputs is at + 1, then the feedback is zero, and if both inputs are at –1, then the output is at –1. For the full electronic implementation of the cell, we will have to make sure that any input in the range (–1, 0) is treated as inactive in the steady state. If we choose $\beta > 1$, then the feedback has gain, β; however, since the cell output is saturated, increasing the gain has the effect of speeding up the transition, and can be used as a global speed control. The constant, –1, can be implemented using the cell bias input.

So far we have only considered the output of cell (i, j) in implementing the row reduction rule. When this cell is activated, then cell $(i, j-1)$ and cell $(i+1, j-1)$ must be deactivated. Applying this operation to each of the input cells to the row reduction rule yields logic equations (3.16) and (3.17).

$$I_z^2(i,j) = I_z^1(i,j) \text{ XOR } (I_z^1(i,j) \text{ AND } I_z^1(i+1,j)) \tag{3.16}$$

$$I_z^2(i,j) = I_z^1(i,j) \text{ XOR } \left(I_z^1(i-1,j) \text{ AND } I_z^1(i,j)\right) \tag{3.17}$$

We note that the AND function, used in each of the above equations, has already been implemented using Equation (3.15), and so we can use the driving signal that actives cell (i, j) to deactivate cell $(i, j - 1)$ and cell $(I + 1, j - 1)$.

The concatenation and overlay rules are provided by logic equations (3.1) and (3.2), respectively. Equation (3.1) can be written as a continuous template using Equation (3.18).

$$A_{i,j}(y_c(t)) = \beta \cdot (y_{i-1,j} + u_{i-1,j} - 1) \tag{3.18}$$

We note here that one of the inputs to the feedback template of Equation (3.18) is not the output of a cell but rather an input that is normally associated with the B_c template. This has removed the need for building a three-dimensional network, which would initially seem the natural approach to resolving the overlay situation where an already activated cell receives another activation signal. Rather than use a separate cell situated above the first cell to hold the second activation, it is, instead, fed back as a time-varying input.

3.4.9 Handling Interference between Cells

We have developed the above design with an implicit assumption that the active cells are isolated in that they will not be affected by parallel operations, with neighboring cells, trying to implement the same operation. In practice, we cannot make this assumption, and so we have to take care of possible interference between cells in the neighborhood. We can imagine the case where a cell, that is, an output for one rule, is also an input for another rule. Since the network is in a continuous dynamic state during the evaluation phase, then the clash of rules could lead to an incorrect steady-state evaluation, or even instability in the network.

An example is provided in [21] that illustrates such an interference situation. Figure 3.12 shows a DBNS map, on the left, that can be reduced by a single application of the row reduction rule. The rule can be applied either on the two leftmost or on the two rightmost nonzero digits. In either application of the rule, we obtain a correct result, as shown on the right of Figure 3.12.

For a CNN implementation of row reduction, every cell has the rule programmed into its templates, and for any cells that are able to activate row reduction (i.e., two adjacent cells in a row), they will attempt to do so. For our example, shown in Figure 3.13(a), the network will attempt simultaneous row reduction on the two leftmost and the two rightmost cells; the result of this is shown in Figure 3.13(b). Since the nonzero digits are able to be reduced once more by the row reduction rule, the network will apply this rule yielding the result shown in Figure 3.13(c), and then enter its steady-state condition. Clearly, however, the result is incorrect, and so we need to program the

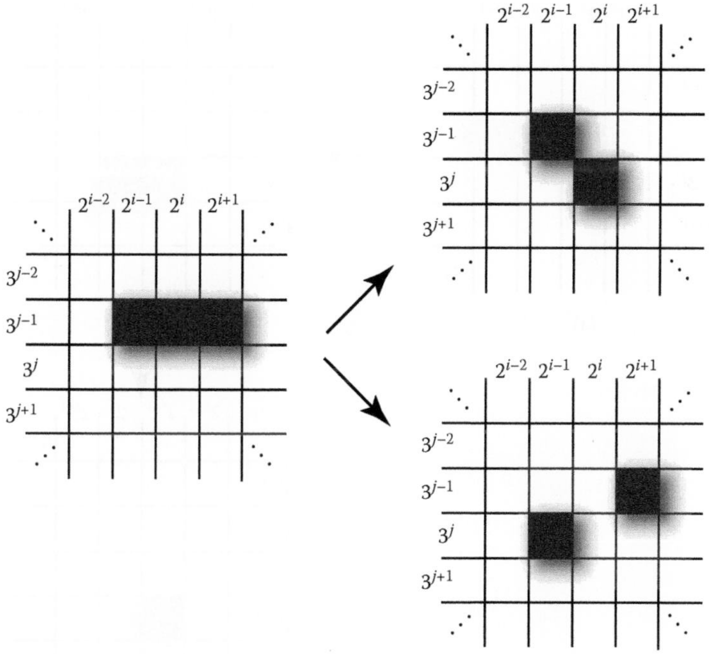

FIGURE 3.12
Correct application of the row reduction rule.

CNN network to recognize this as a special case and avoid the simultaneous reduction.

In Figure 3.14, we show two other cases of failure in concurrent applications of row reduction. The first, shown in Figure 3.14(a), shows a conflict at cell (i, j) which is active prior to the concurrent application of the rule to cells $(i, j-1)$, $(i+1, j-1)$ and to cells $(i-1, j)$, (i, j). Application of the rule to the first two cells will attempt to form an overlay at cell (i, j), whereas the application of the rule to the second group of cells will attempt to deactivate cell (i, j). This constitutes the equivalent of a race condition that is well known in combinational logic. Another example, in Figure 3.14(b), initiates a conflict in cell $(i-1, j)$, which again faces a competing condition of an overlay (from row reduction of cells $(i-1, j-1)$, $(i, j-1)$) and deactivation (from row reduction of cells $(i-1, j)$, (i, j)).

To avoid such conflicts, the following two restrictions are necessary [21]:

1. At any time, an active cell can only participate in one reduction rule.
2. At any time, the application of a reduction rule on any group of active cells cannot affect an active cell that is a candidate for another reduction rule.

Since we need to prioritize the selection of a reduction rule when there is a conflict, the following theorem provides the best priority ordering [21].

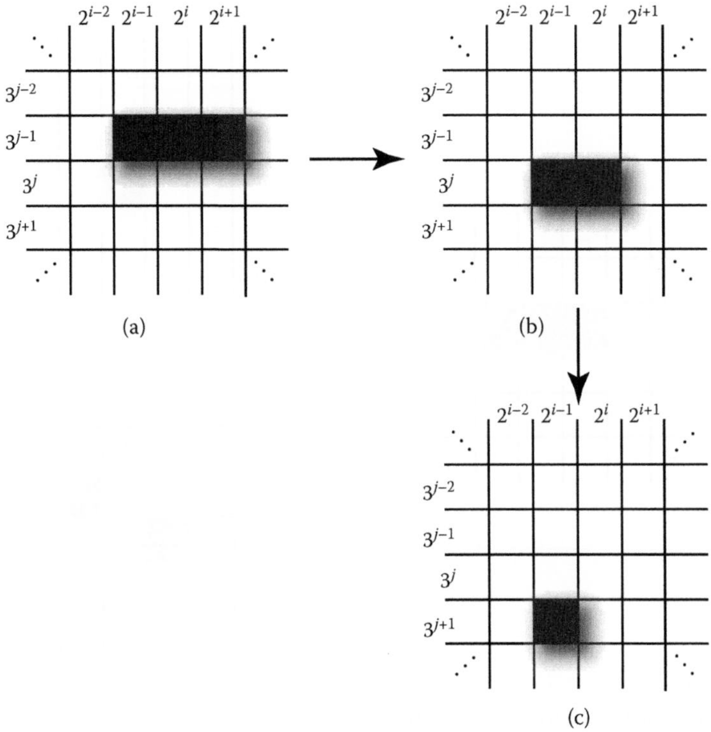

FIGURE 3.13
Failure of CNN implementation of concurrent row reduction rules.

Theorem 3.1: Given any DBNR map, applying the reduction rules to active groups in positions (i, j) in descending order of i and j results in a DBNR map with fewer active cells than with any other ordering.

An example of the application of this priority ordering and reverse ordering (for comparison) is shown in Figures 3.15 and 3.16. In both cases the rows were scanned for reduction rule applications in descending order of j. For Figure 3.15 the columns were scanned for reduction rule application in descending order of i, and for Figure 3.16 the columns were scanned in ascending order for i. Note that descending order of i corresponds to the best priority ordering from Theorem 3.1.

3.4.10 CMOS Implementation of DBNS CNN Reduction Rules

As with analog computers, the use of current as a variable in CMOS circuit implementations is a preferred solution. Kirchoff's current law is much easier to use in addition and subtraction than Kirchoff's voltage law. A single junction of multiple input current sources acts as a summation unit on the

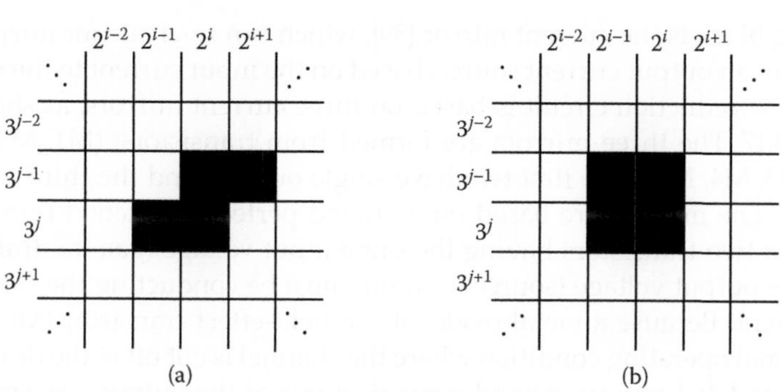

FIGURE 3.14
Two other special cases of failure in concurrent row reduction.

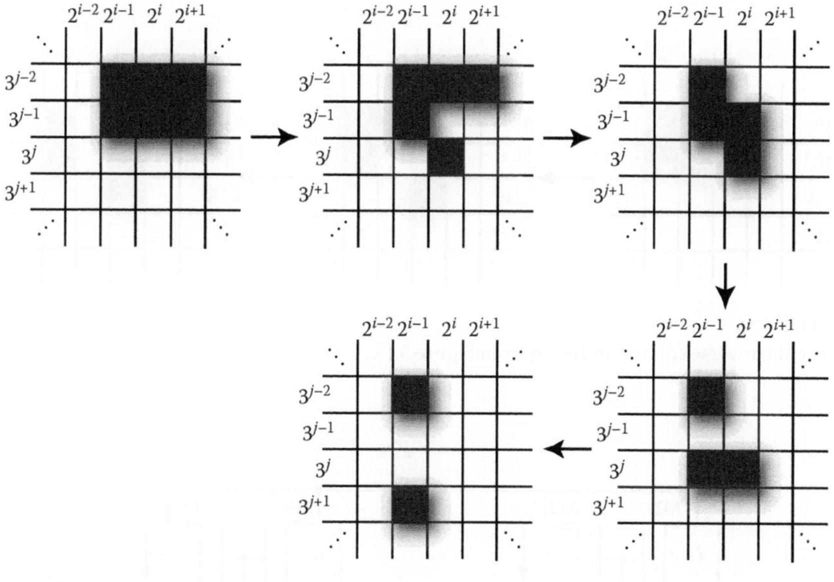

FIGURE 3.15
An example of priority ordering of reduction [21].

output current; by changing the direction of current sources, we can perform addition and subtraction with this single junction.

In analog computers, the linear accuracy is provided by very accurate passive components (resistors and capacitors) with virtual grounds as summing junctions, a virtual ground being the result of negative feedback across a very high-gain amplifier (the amplifiers are carefully designed to avoid oscillations when feedback is applied).

In CMOS-integrated analog circuits, naturally close matching between adjacent integrated transistors is used to reflect and steer currents. A basic building block is the current mirror [39], which can re-create (or mirror) the current in an output current source, based on the input current to the mirror.

The row reduction circuit is based on three current mirrors, as shown in Figure 3.17. The three mirrors are formed from transistors: (M1, M2), (M6, M7), (M3, M4, M5). The first two have single outputs and the third has two outputs. The mirrors are based on assumed perfectly matched transistors, such that two transistors having the same input voltage (gate to drain) and the same output voltage (source to drain) must be conducting the same output current. Because a metal-oxide-silicon field-effect transistor (MOSFET), in a normal operating condition where the channel is cut off at the drain end, can be modeled as a very good current source at the output, we can allow different sources to drain voltages on the two transistors and still obtain quite close matching. If this is not sufficient, then feedback can be applied (such as using matched resistors in series with the drains of each transistor) to improve the matching characteristics [39]—in a sense, using matched

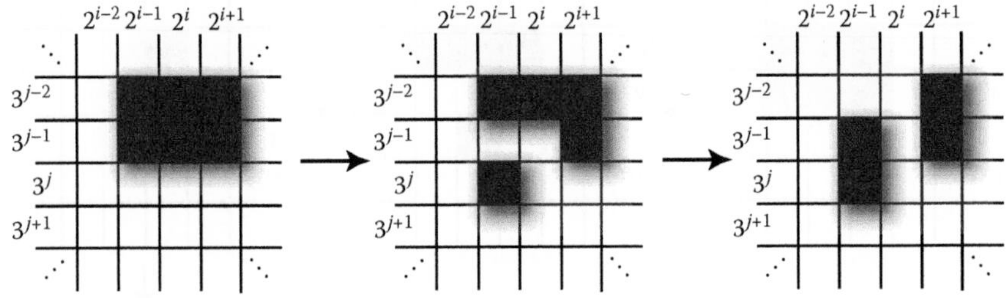

FIGURE 3.16
Using the reverse column ordering from Figure 3.15.

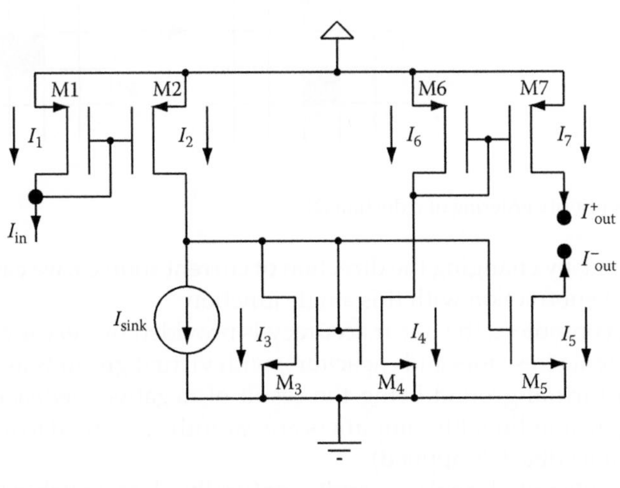

FIGURE 3.17
A CMOS implementation of the row reduction rule.

passive components with feedback to improve linearity, as in the application of operational amplifiers with feedback in analog computers. We also assume that the input to the gate is an open circuit (no current flow into the gate). This is quite a good approximation at DC.

The first current mirror in Figure 3.17, (M1, M2), uses *p*-channel transistors that are driven from negative source-drain and gate-drain voltages. The drain voltages are at the same potential (positive supply) and the gate voltages are connected together, so the gate-drain voltages are the same: $V_{GS1} = V_{GS2}$. Although the source voltages are at different potentials, we will assume that the currents match (as discussed above). The input current, I_3, to the (M3, M4, M5) mirror is given by Equation (3.19), and this current is mirrored to both I_{out}^+ and I_{out}^-.

$$I_3 = \begin{bmatrix} I_{in} - I_{sink} & \text{for } I_{in} \geq I_{sink} \\ 0 & \text{for } I_{in} < I_{sink} \end{bmatrix} \tag{3.19}$$

The connection of the row reduction template to the three-cell configuration is shown in Figure 3.18. The template uses the sum of the output currents from cell (i, j) and cell $(i + 1, j)$ to drive the output currents I_{out}^+ and I_{out}^-. These currents, in turn, are used to deactivate the cells and to activate cell $(i, j + 1)$. For now we will ignore the obvious race condition associated with this feedback. Providing that $I_{i,j} + I_{i+1,j} > I_{sink}$ and $(I_{i,j}, I_{i+1,j}) < I_{sink}$, the upper condition in Equation (3.19) will only be met if both input cells are active, i.e., a valid row reduction rule application.

3.4.11 A Complete CNN Adder Cell

The circuit developed in the previous section assumes that the three-cell row reduction structure is isolated from the rest of the network, so that we

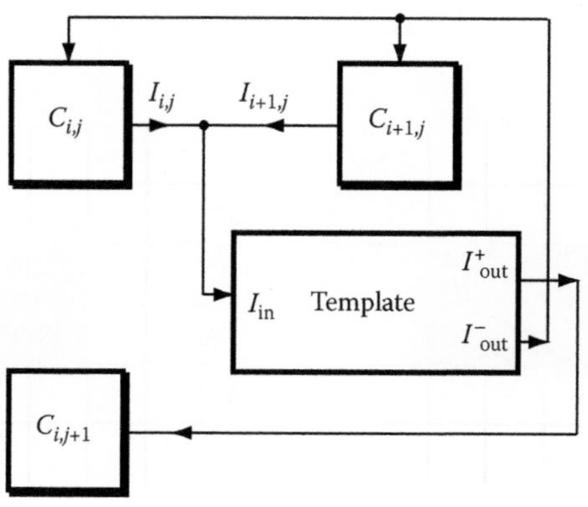

FIGURE 3.18
Connecting cells to the row reduction template.

do not have to be bothered with problems associated with ordering of reduction rules, or incorrect operation resulting from interactions of competing rule applications. These problems are associated with the CNN equivalent of race conditions in combinational logic. An important observation to be made is that if we are able to enforce the ordering of the application of the reduction rule, then that will take care of the problem of competing concurrent applications of the rule [21]. Figure 3.19 illustrates the conflict with overlapping groups of active row reduction cells (this is the same example used in Figures 3.15 and 3.16). From Theorem 3.1, the four overlapping groups should have a priority ordering of activation of G1 $\Rightarrow$ G2 $\Rightarrow$ G3 $\Rightarrow$ G4, with the top priority of G1. This ordering can be forced by using the deactivation current mirror output I_{out}^{-}, as shown in Figure 3.17, as feedback to temporarily halt the application of the reduction rule until higher-priority cells have completed their reduction rule activation (where $I_{out}^{-} = 0$ for all the higher-priority cell groups). As an example, the connections required for the G4 cell are shown in Figure 3.20 [21]. A complete adder cell schematic is shown in Figure 3.21; the cell includes the basic CNN circuitry from Figure 3.17 along with row reduction and priority ordering circuitry and overlay rule circuitry.

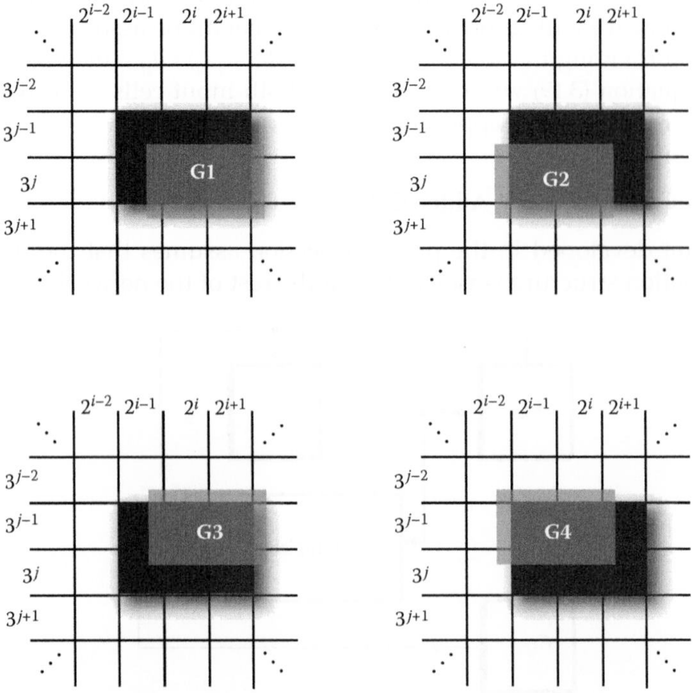

FIGURE 3.19
Overlapping cell groups.

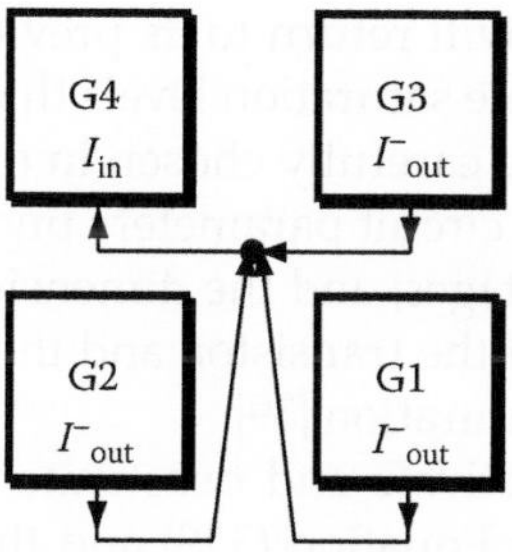

FIGURE 3.20
Applying feedback to control priority ordering to the row reduction groups.

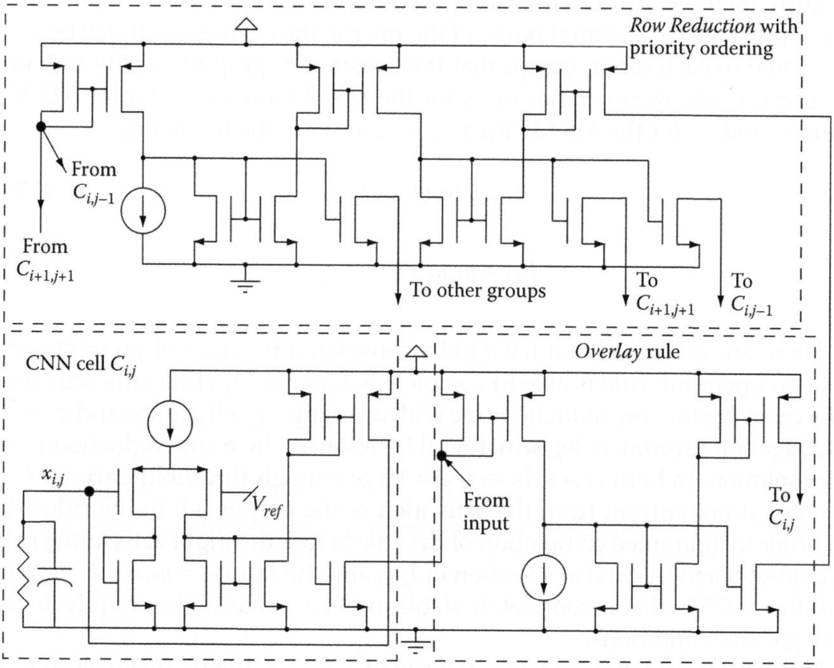

FIGURE 3.21
CNN adder cell schematic[21].

3.4.12 Avoiding Feedback Races

In Section 3.4.10 we introduced the idea of feeding back output cell activation information to the input of the basic CNN cell in order to implement reduction rules. It is clear that this action could cause a race condition where the changing output of a cell, which is dependent on specific input values, could at the same time change the very inputs that are driving the activation function. This is similar to a flip-flop circuit where a transient input value has to reach a certain level before the circuit switches state. If this level is not reached, then the flip-flop will return to its previous stable state. The CNN network also has steady-state saturation levels that are driven by inputs and offset biases that have to be carefully chosen in order to guarantee successful (correct) operation. The circuit parameters under our control are bias, or reference, currents and voltages, and the dimensions of the transistor channels that control the gain of the transistor, and the effective resistance of the channel when it is not in saturation [39].

We can determine the activate and deactivate output currents of Figure 3.17, I_{out}^{+} and I_{out}^{-}, by using Equation (3.19) and the gain of the current mirrors. In Section 3.4.10 we assumed that current mirrors reflected the same current at the output as the input current. If we use transistors in the current mirror with different channel dimensions (i.e., different ratios of width to length), then we will mirror either a larger or smaller current than enters the input. Over the normal range of the mirror, the currents will still be proportional to each other, except that the constant of proportionality will not be unity. Using mirror ratios of m_1 for the M3-M5 mirror, m_2 for the M3-M4 mirror, and m_3 for the M6-M7 mirror, we can write the following:

$$I_{out}^{-} = m_1(I_{in} - I_{sink}) \tag{3.20}$$

$$I_{out}^{+} = m_2 m_3(I_{in} - I_{sink}) \tag{3.21}$$

There are two cases that have to be considered in terms of guaranteeing that an operating rule is able to complete before a competing rule activates. The cases are the row reduction rule with overlapping cell groups and a carry propagation (overlay rule), which will be followed by a row reduction rule. The solution, in both cases, is to use a large enough threshold current, I_{sink}, on the output current from the activation of the output cell in the rule [21], in order to guarantee completion of the rule before the input cells to the rule are deactivated. A careful selection of I_{sink}, and the ratio $m = m_1 = m_2 \cdot m_3$ will produce a CNN that is completely stable in steady state, with no equivalence to logic race conditions.

Extensive simulations of the self-timed CNN network demonstrate its feasibility for reducing DBNS maps with overlapping reduction rule cell groups. An example of the reduction process, involving both row reduction and overlay rules, is shown in Figure 3.22. The upper part of the figure shows the reduction process on the DBNS map, following the optimum priority ordering as given in Theorem 3.1. The black and white squares indicate saturated values at + 1 and 0, respectively; the gray squares indicate intermediate values between the two saturation levels. The lower part of the diagram shows the output waveform of the cell corresponding to each square in the DBNS map. The waveforms were obtained from HSPICE simulations, based on integrated circuit transistor models and parasitic values for a typical layout (in this case based on the TSMC 180 nm technology) [21].

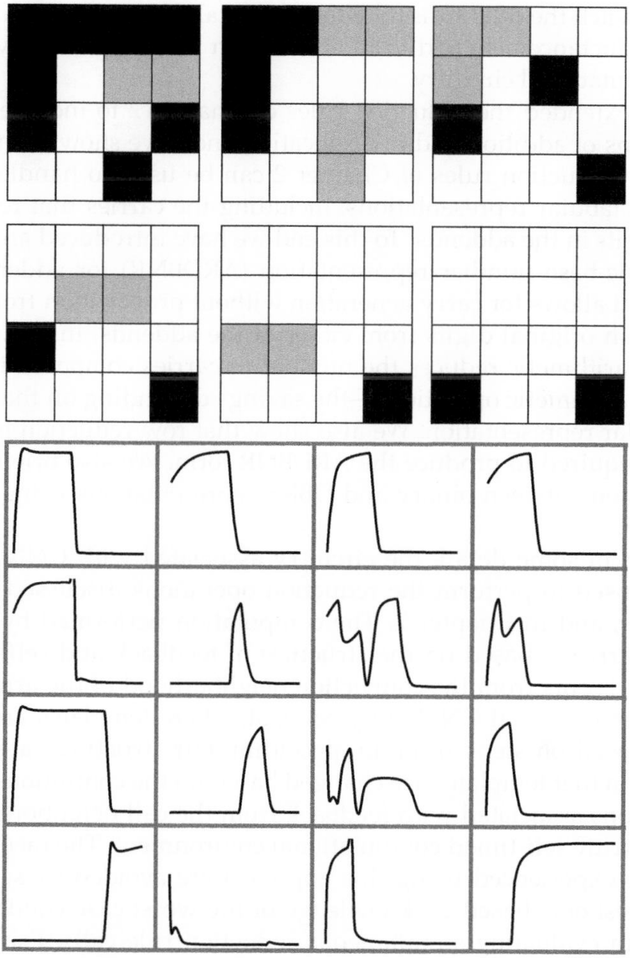

FIGURE 3.22
HSPICE simulation of prioritized reduction rules.

3.5 Summary

In this chapter we have introduced the first implementation circuitry in the book, namely, the use of cellular neural networks (CNNs) to perform arithmetic based on the DBNS. Our motivation is based on the similarities between the symbolic substitution reduction rules introduced in Chapter 2 and many image processing algorithms that operate on a neighborhood of pixels. The successful use of CNN architectures to compute such algorithms leads to their potential success in implementing the reduction rules of DBNS representations. The advantages of CNN architectures that are embodied in the asynchronous behavior of the arrays and the ability to smoothly switch the cells so as to reduce cross talk and voltage drops across parasitic inductance also add to the interest in using CNN arrays in very dense computational circuitry.

We have extended the reduction rules of Chapter 2 to include arithmetic operations of addition and multiplication and have shown that the row and overlay reduction rules of Chapter 2 can be used to handle addition with DBNS tabular representations, including the carries that result from overlaid digits in the addends. To this end we have introduced an addition-ready double-base number representation (ARDBNR) for addends in an addition that allows for carry generation without propagation from further overlays with original digits from either of the addends. In fact, we show that DBNS arithmetic reduces the number of carries compared to equivalent binary arithmetic operations—the savings depending on the structure of the tabular representation. We also show that row reduction is the only operation required to produce the ARDBNR form. We also briefly discuss the conversion between binary and DBNS representations using symbolic substitution.

We cover, in some detail, the circuitry associated with CNN cells and the arrays used to perform the reduction operations discussed earlier in this chapter, and in Chapter 2. The computation performed by each cell in a CNN array is based on the structure of feedback and self-templates that receive inputs from the connection neighborhood of the array. Rather than use the universal CNN array concept, where templates are globally switched based on steps from an algorithm, our structures are self-programming in that templates are changed based on the continuously changing conditions associated with feedback from the call neighborhood. This provides a truly self-timed computational environment. The race problems that can be experienced using this approach are avoided by setting bias current thresholds based on knowledge of the worst-case conditions that can occur in evaluating simultaneous reduction rule calls. Examples are provided to illustrate the conflict avoidance, and simulations of the circuits developed in the chapter are used to demonstrate the operation of relatively large arrays.

References

1. V.S. Dimitrov, T.V. Cooklev, Hybrid Algorithm for the Computation of the Matrix Polynomial, *IEEE Transactions on Circuits and Systems*, 42(7), 377–380, 1995.
2. V.S. Dimitrov, T.V. Cooklev, Two Algorithms for Modular Exponentiation Using Nonstandard Arithmetic, *IEICE Transactions on Fundamentals*, E78-A, 82–87, 1995.
3. L.O. Chua, L. Yang, Cellular Neural Networks: Theory, *IEEE Transactions on Circuits and Systems*, CAS-35, 1257–1272, 1988.

4. T. Roska, L. Chua, The CNN Universal Machine: An Analogic Array Computer, *IEEE Transactions on Circuits and Systems–II*, 40(3), 163–172, 1993.
5. Á. Zarándy, A. Stoffels, T. Roska, L.O. Chua, Implementation of Binary and Gray-Scale Morphology on the CNN Universal Machine, *IEEE Transactions on Circuits and Systems*, 45, 163–167, 1998.
6. M. Pankaala, A. Paasio, M. Laiho, Implementation Alternatives of a DBNS Adder, in *9th International Workshop on Cellular Neural Networks and Their Applications*, May 28–30, 2005, pp. 138–141.
7. A.J. Martin, M. Nyström, Asynchronous Techniques for System-on-Chip Design, *Proceedings of the IEEE* (Special Issue on Systems-on-Chip), 94(6), 1089–1120, 2006.
8. E.G. Friedman, *High Performance Clock Distribution Networks*, Kluwer Academic, Norwell, MA, 1997.
9. J.J. Becerra, E.G. Friedman, *Analog Design Issues in Digital VLSI Circuits and Systems*, Kluwer Academic, Norwell, MA, 1997.
10. M.D. Ercegovac, T. Lang, *Digital Arithmetic*, Morgan Kaufmann (Elsevier Science), San Francisco, 2004.
11. T. Bose, Z. Zhang, O. Chauhan, M. Radenkovic, Design of Multiplier-Free State-Space Digital Filters, *IEEE International Symposium on Circuits and Systems*, 3, 2591–2594, 2005.
12. L.O. Chua, L. Yang, Cellular Neural Networks: Applications, *IEEE Transactions on Circuits and Systems*, CAS-35, 1273–1289, 1988.
13. L.O. Chua, T. Roska, The CNN Paradigm, *IEEE Transactions on Circuits and Systems*, CAS-40, 147–155, 1993.
14. T. Matsumoto, L.O. Chua, H. Suzuki, CNN Cloning Template: Connected Component Detector, *IEEE Transactions on Circuits and Systems*, 37(5), 633–635, 1990.
15. T. Matsumoto, L.O. Chua, R. Furukawa, CNN Cloning Template: Hole-Filler, *IEEE Transactions on Circuits and Systems*, 37(5), 635–638, 1990.
16. S. Sadeghi-Emamchaie, Novel Cellular Neural Networks for Low-Noise Digital Arithmetic Architectures, PhD thesis, University of Windsor, Ontario, Canada, 1999.
17. S. Sadeghi-Emamchaie, G.A. Jullien, W.C. Miller, Very Low-Noise (Switching Free) CNN-Based Adder, *SPIE Proceedings in Advanced Signal Processing Algorithms, Architectures, and Implementations IX*, 3807(7), 1999.
18. Y. Ibrahim, G.A. Jullien, W.C. Miller, Ultra Low Noise Signed Digit Arithmetic Using Cellular Neural Networks, in *Proceedings of 4th International Workshop*, 2004, pp. 136–142.
19. Y. Ibrahim, G.A. Jullien, W.C. Miller, Arithmetic Implementation Techniques Using Analog Cellular Neural Networks, presented at 17th IMACS World Congress on Scientific Computation, Modeling and Applied Mathematics, IMACS '05, 2005.
20. Y. Ibrahim, G.A. Jullien, W.C. Miller, V.S. Dimitrov, DBNS Addition Using Cellular Neural Networks, in *IEEE International Symposium on Circuits and Systems*, ISCAS '05, 2005, pp. 3914–3917.
21. Y. Ibrahim, Novel Arithmetic Implementations Using Cellular Neural Network Arrays, PhD thesis, University of Windsor, Ontario, Canada, 2005.
22. H.T. Kung, Why Systolic Architectures, *IEEE Computer Magazine*, 15(1), 37–46, 1982.

23. B. Christie, The MA7170 Systolic Correlator, Architecture and Applications, in *Systolic Arrays*, a collection of papers presented at the First International Workshop on Systolic Arrays, Oxford, W. Moore, A. McCabe, R. Urquhart (eds.), Adam Hilger, Bristol, 1986.
24. J.V. McCanny, J.G. McWhirter, Completely Iterative, Pipelined Multiplier Array Suitable for VLSI, *IEE Proceedings*, 129(Pt. G, no. 2), 40–46, 1982.
25. J.V. McCanny, J.G. McWhirter, Implementation of Signal Processing Functions Using 1-Bit Systolic Arrays, *Electronic Letters*, 18(6), 241–243, 1982.
26. J.V. McCanny, J.G. McWhirter, Bit-Level Systolic Array Circuit for Matrix Vector Multiplication, *IEE Proceedings*, 130(Pt. G, no. 4), 125–130, 1983.
27. J.V. McCanny, J.G. McWhirter, Optimised Bit Level Systolic Array for Convolution, *IEE Proceedings*, 131(Pt. G, no. 6), 632–637, 1984.
28. S.Y. Kung, *VLSI Array Processors*, Prentice-Hall, Englewood Cliffs, NJ, 1988.
29. L. Török, Stability, Optical Flow and Stochastic Resonance in Cellular Wave Computing, PhD thesis, Pétér Pázmány Catholic University, Hungary, November 2005.
30. E.G. Friedman, Clock Distribution Networks in Synchronous Digital Integrated Circuits, *Proceedings of the IEEE*, 89(5), 665–692, 2001.
31. H. Zarrabi, Z. Zilic, Y. Savaria, A.J. Al-Khalili, *On the Efficient Design and Synthesis of Differential Clock Distribution Networks*, Z. Wang (ed.), Intech, Rijeka, Croatia, VLSI 2010, pp. 331–352.
32. D. Atienza, F. Angiolini, S. Murali, A. Pullini, L. Benini, G. De Micheli, Network-on-Chip Design and Synthesis Outlook, *Integration, the VLSI Journal*, 41, 340–359, 2008.
33. P. Larsson, Power Supply Noise in Future IC's: A Crystal Ball Reading, in *Proceedings of IEEE Custom Integrated Circuits Conference*, 1999, pp. 467–474.
34. J. Martin, M. Nyström, Asynchronous Techniques for System-on-Chip Design, *IEEE Proceedings* (Special Issue on SoC), 4(6), 1089–1120, 2006.
35. S. Espejo, R. Dominguez-Castro, G. Liñán, Á. Rodríguez-Vázquez, A 64 × 64 CNN Universal Chip with Analog and Digital I/O, in *Proceedings of 5th IEEE International Conference on Electronics, Circuits and Systems*, ICECS '98, Lisbon, Portugal, 1998, pp. 203–206.
36. H. Kim, T. Roska, H. Son, I. Petras, Analog Addition/Subtraction on the CNN-UM Chip with Short-Time Superimposition of Input Signals, *IEEE Transactions on Circuits and Systems–I*, 50(3), 429–432, 2003.
37. S. Sadeghi-Emamchaie, Novel Cellular Neural Networks for Low-Noise Digital Arithmetic Architectures, PhD thesis, University of Windsor, Ontario, Canada, 1999.
38. Z. Galias, Designing Cellular Neural Networks for the Evaluation of Local Boolean Functions, *IEEE Transactions on Circuits and Systems-II*, 40, 219–223, 1993.
39. A. Sedra, K. Smith, *Microelectronic Circuits*, 3rd ed., Saunders College Publishing, Philadelphia (a division of Holt, Rinehart and Winston), 1991.

4

Multiplier Design Based on DBNS

4.1 Introduction

The problem of multiplying two integers is of such paramount importance that one does not need specific references to justify the necessity to investigate it as deeply as possible. However, for researchers outside the computer arithmetic community, it seems rather surprising to realize that this simple computational problem is still not thoroughly understood and, more to the point, that there is still room for improvements in both theory and practice. Part of this chapter is based on reference [42] (© IEEE 2011).

4.1.1 A Brief Background

In the mid-1950s Kolmogorov [1] made a conjecture that any multiplication algorithm will require $\Omega(k^2)$ elementary bit operations, where k is the binary length of the multiplicands. This conjecture was disproved in a constructive manner by Karatsuba [2–4], who proposed an algorithm that uses $O(k^{1.585})$ elementary bit operations. Karatsuba's multiplication algorithm has led to the discovery of many similar algorithms; perhaps, the most famous example is Strassen's matrix multiplication algorithm [5], which effectively demonstrated that one can multiply two $n \times n$ matrices by using $o(n^3)$ multiplications and additions. Researchers in the 1960s tried to improve, asymptotically, Karatsuba's multiplication algorithm and they succeeded. In 1971 Schönhage and Strassen published an algorithm [6] with asymptotic complexity $O(k \log k \operatorname{loglog} k)$, that is, demonstrating that the multiplication can be performed in nearly linear time. For almost 40 years this result has not been improved; however, in 2007 Fürer [7] designed an algorithm with lower asymptotic complexity, namely, $O(k \log k \operatorname{loglog} \cdots \log k)$ operations. Whether it is possible to multiply two numbers in purely linear time is still an open problem. Under some restrictive assumptions it is possible to prove a lower bound of $\Omega(k \log k)$ elementary bit operations [8].

4.1.2 Extremely Large Numbers

The above-mentioned algorithms (Karatsuba's, Schönhage-Strassen's, and Fürer's) all have a subquadratic complexity. However, the implicit constant associated with the big-O notation is very large, and this severely limits their applicability to problems of practical importance. Karatsuba's multiplication outperforms the classical shift-and-add algorithm if the size of the multiplicands is around 1,000 bits. This makes it suitable for specific cryptographic applications. The algorithms by Schönhage-Strassen and Fürer are useful if one deals with extremely large numbers. Applications include computational number theory and computations associated with the search of large Mersenne primes and finding divisors of Fermat numbers. In those cases we deal with numbers having more than 1 million decimal digits.

4.2 Multiplication by a Constant Multiplier

If one of the operands is a constant, it is possible to perform optimizations for the series of additions or subtractions required to perform the multiplication. Lefévre [9] was the first to prove a lower bound on the number of additions/subtractions required in implementing multiplication by k-bit constants.

Theorem 4.1: Multiplication by a k-bit constant requires $O(k/\log k)$ additions/subtractions in the worst case [9].

In 1996 Pinch [10] proved that one can implement multiplication by a k-bit number by using $O(k/(\log k)^{\alpha})$ additions/subtractions, where $\alpha < 1$.

The above two results are based on the idea of eliminating, as much as possible, any common subexpressions in the signed digit representation of the constant and to reuse, as much as possible, any intermediate results obtained.

In 2007 a stronger result was proved [11], namely, that the condition $\alpha < 1$ in Pinch's theorem is not necessary and, in fact, multiplication by any k-bit constant can be achieved by using $O(k/\log k)$ additions and shifts.

Theorem 4.2: Multiplication by a k-bit constant needs no more than $O(k/\log k)$ additions in the worst case [11,12].

The fact that multiplication by a k-bit constant can be performed in $O(k/\log k)$ additions, where one addition requires $O(k)$ elementary bit operations, suggests that perhaps it would be possible to design an algorithm with asymptotic complexity $O(k^2/\log k)$ elementary bit operations, which is $o(k^2)$. The constant multiplication algorithms suggested in [11] have two very attractive features: first, their complexity estimate has a very small implicit constant, and second, the conversion of the constant can be performed very quickly. In fact, it is so fast that it does make sense to try a similar approach to the general multiplication problem.

4.3 Using the DBNS

Theorem 4.2, that one only need use $O(k/\log k)$ additions in implementing the multiplication by any k-bit number, is based on a specific double-base representation of the constant. In this case we use the general definition for the double-base number system (DBNS) (repeated below from Chapter 2, Definition 2.1).

Definition 4.1: Given p, q, two different prime numbers, the double-base number system (DBNS) is a representation scheme in which every positive integer x is represented as the sum or difference of $\{p, q\}$-integers, i.e.,

$$x = \sum_{i=1}^{l} s_i p^{a_i} q^{b_i} \text{ with } s_i \in \{-1,1\} \text{ and } a_i, b_i \geq 1 \quad (4.1)$$

In finding the canonic DBNS representations in a reasonable amount of time, especially for large integers (e.g., cryptographic size numbers), we use the greedy algorithm (Chapter 2, Section 2.4) to find a fairly sparse representation very quickly.

The most important feature of the greedy algorithm is that it guarantees an expansion of sublinear length, where, from Theorem 4.1, the greedy algorithm terminates after $K = O(\log x/\log\log x)$ steps.

If one applied the greedy algorithm without any modifications, the largest power of 3 that might occur in the DBNS representation of x is $\log_3 x = O(\log x)$. If one encodes the integers in the way shown in the DBNS tabular representations from Chapters 2 and 3, then one will need rather large tables, which could be a drawback in many applications. The following theorem shows that even if we use very few powers of 3, we can still achieve a DBNS representation of x having sublinear length.

Theorem 4.3: Every positive integer, x, can be represented as the sum of at most $O(\log x/\log\log x)$ $\{2,3\}$-integers such that the largest power of 3 is bounded by $O(\log x/\log\log x)$.

Proof: Consider the binary representation of x (which contains $\log x$ bits). Now, we break down this representation into $\log\log x$ blocks of bits, each of length $O(\log x/\log\log x)$. Then, according to Theorem 4.2, every block can be represented as the sum of at most

$$O\left(\frac{\dfrac{\log x}{\log\log x}}{\log\left(\dfrac{\log x}{\log\log x}\right)}\right) = O\left(\frac{\log x}{\log^2 \log x - \log\log x \log\log\log x}\right)$$

$$= O\left(\frac{\log x}{\log^2 \log x}\right)$$

{2, 3}-integers. As the number of blocks is loglog x, we conclude that this representation consists of, at most, O (log x/loglog x) {2, 3}-integers. The highest power of 3 that occurs in such a representation is governed by the highest power of 3 that can occur in every block, and it is at most O (log x/loglog x). ∎

Theorem 4.2 is constructive and leads to the simple algorithm of Figure 4.1 for computing a DBNS representation of length O (log x/loglog x) with restricted powers of 3.

Theorem 4.3 and the greedy algorithm (from Chapter 2, Section 2.4) are the key ingredients in proving Theorem 4.2, that is, the fact that multiplication by a k-bit constant needs no more than $O(k/\log k)$ additions in the worst case. If one uses the standard binary expansion of a k-bit constant, then the number of additions is one less than the Hamming weight of the constant, that is, $O(k)$ if one uses a standard binary or a signed digit binary representation

Input: Positive integer x; block size w, precomputed canonic representations of every integer:

$$\sum_{i=0}^{w-1} = d_i 2^i, d_i \in \{0,1\}$$

Output: List, **L**, of {2, 3}-integers representing the DBNS expansion of x with powers of 3 restricted by

$$O\left(\frac{\log n}{\log\log n}\right)$$

```
procedure      blocking (x)
{
L={}
for j = 0 to ⌈k/w⌉ do
{
Process block of length w;
Find canonic representation of Σ_{i=0}^{w-1} d_{i+jw} 2^i from the
precomputed table;
Multiply each term of the expansion by 2^{jw} and add to L;
}
end for;
}
```

FIGURE 4.1

Blocking algorithm for computing a DBNS expansion of x. (From V.S. Dimitrov, K.U. Jarvinen, J. Adikari, Area-Efficient Multipliers Based on Multiple-Radix Representations, *IEEE Transactions on Computers*, 60(2), 189–201, 2011. Copyright © 2011. With permission.)

of the constant. However, if the constant is represented in DBNS format, then the number of additions will be equal to the number of {2, 3}-integers necessary to represent this constant plus the highest power of 3 that occurs in the corresponding DBNS representation (because the multiplication by 3 needs one shift and one addition). According to Theorem 4.2, if one restricts the highest power of 3 to $O(k/\log k)$, it is still sufficient to guarantee that the DBNS expansion found by the blocking algorithm will be of length $O(k/\log k)$.

Therefore, the total number of additions will be bounded from above by $O(k/\log k)$.

The example provided below demonstrates the power of the above results:

Example 4.1

Show that the multiplication by any 300-bit constant can be achieved by using at most 77 additions.

Let us consider the binary representation of a 300-bit number and let us break this representation into ten 30-bit blocks. According to Table 4.1, every 30-bit integer can be represented by using at most six {2, 3}-integers (because $2^{30} < 1{,}441{,}896{,}119 < 2^{31}$). Since $31^8 < 2^{30} < 3^{19}$, the execution of the algorithm in Figure 4.1 will return a DBNS expansion of any 300-bit number of length at most 10 (the number of 30-bit blocks) × 6 (the maximal number of {2, 3}-integers per block) = 60. The highest power of 3 that might occur in this expansion is 18; therefore in the worst case, one will need 60 + 18 − 1 = 77 additions.

It is interesting to note that none of the available methods on constant multiplication lead to such concrete, easy-to-understand and interpret, and nontrivial bounds. In the literature it is very often pointed out that the only provable upper bound, even if one is allowed to use subtractions, is $k/2$, and it can only be achieved if one uses the so-called nonadjacent form (NAF) representation for the constant.

The above example with 300-bit numbers deserves a few comments. It is clear that the block size that we have selected, 30 bits, would have been somewhat large, especially for a hardware implementation. But even if one uses much smaller blocks, say 12-bit blocks—a completely appropriate size from both a software and a hardware point of view—the estimate that can be found is still useful. In the worst case one will need 25 × 4 (all 12-bit numbers can be represented by using at most four terms; see Table 2.1 in Chapter 2) plus 7 (the highest possible power of 3) minus 1, that is, 106. In this case one can appreciate the savings that can be achieved if we allow subtractions. Up to 4,984 we can represent any number as the sum or difference of at most three {2, 3}-integers. Therefore if we can use subtractions, then the algorithm with small 12-bit blocks will require in the worst case 25 × 3 + 7 = 81 additions/subtractions. If one inspects all the canonic representations up to 4,095, one will see that the highest power of 2 that occurs is only 12; therefore the risk of overflow is completely under control.

TABLE 4.1

Encoding Scheme for 7-Bit Numbers

	Digit Selection	Binary Exponent	Digit Selection	Sign	Binary Exponent
7-bit number	2 bits	3 bits	2 bits	1 bit	3 bits

Source: V.S. Dimitrov, K.U. Jarvinen, J. Adikari, Area-Efficient Multipliers Based on Multiple-Radix Representations, *IEEE Transactions on Computers*, 60(2), 189–201, 2011. Copyright © 2011. With permission.

Since the introduction of the DBNS in 1995 [16] it has found various applications in digital signal processing and cryptography. Initial applications in the field of digital signal processing [17–23] have shown that the DBNS, if properly applied, can lead to very considerable improvements over standard binary number system designs. Since the first publication, a number of fabricated designs [17,19,20] have demonstrated that this number system is ideally suited for inner product computations and often outperforms the equivalent binary design by a considerable margin in terms of every important figure of merit. There is also ongoing research on a general purpose processor that implements all of the arithmetic operations over DBNS. It will be interesting to see the outcome of this project.

Since 2005, researchers in the area of public key cryptography have discovered that the attractive features of the DBNS make it extremely well suited for cryptographic applications, particularly in implementing point and multiple-point multiplications over elliptic curves. The number of references grows very rapidly. The reader will find useful information in [13,14,16, 24–38].

This published research work shows that this number representation can be successfully used for, perhaps, the most fundamental computational operation: multiplication of two integers.

4.4 DBNS Multiplication with Subquadratic Complexity

From the previous section we have seen that once we have converted one of the multiplicands (k-bit integer) into a DBNS format with a restricted power of 3, the final part of the multiplication algorithm will require $O(k^2/\log k)$ elementary bit operations. So, the only thing that remains to be analyzed is the cost of conversion.

For the conversion task we have essentially two options: First, we can consider the loop-up table-based approach. It requires the search over each of the precomputed $O(\log k)$ blocks, each block corresponding to a number of size $O(k/\log k)$. The entire task will require $(\log^2 k)$ elementary bit operations, so the overall complexity will be $O((k^2/\log k)+\log^2 k)=O(k^2/\log k)$.

Second, we have to consider the transformation of the multiplicand on the fly by using a memory-free algorithm with low computational complexity. The algorithm proposed by Berthé and Imbert allows a very fast implementation of step 3 of the greedy algorithm [29]. For a k-bit integer it computes the closest {2, 3}-integer in $O(\log k)$ operations, each involving $O(k)$ elementary bit operations. If this algorithm is used, then one will need $O(k \log k)$ elementary bit operations to convert the entire multiplicand into the desired DBNS format. Therefore, the total bit complexity of the multiplication procedure would remain $O(k^2/\log k)$. This can be encapsulated as:

Theorem 4.4: The DBNS multiplication algorithm requires $O(k^2/\log k)$ elementary bit operations.

As we have pointed out, there are several other subquadratic multiplication algorithms, and they all have a superior asymptotic behavior compared to the algorithm described here. However, they all suffer from the fact that the implicit constant associated with their complexity analysis is very large. This drawback severely limits their applicability, as discussed in the introduction to this chapter.

What can we say about the constant hidden in the $O(k^2/\log k)$ estimate of our algorithm? It depends entirely on the constant associated with the complexity of the greedy algorithm (Theorem 4.2). That is, for a given k-bit number, one anticipates $C(k/\log k)$ {2, 3}-integers in its DBNS representation. What do we know about the constant C? There is a big gap between computational experiments and what the best available theory can rigorously and unconditionally prove. The only thing that the theory can guarantee is that this constant is less than $\log(C(r))$, where $C(r)$ is the constant in Baker's theorem [15]. That would mean that we can only prove that C is less than 10^7. The computational experiments very strongly suggest that C is very close to unity, and under certain reasonable probabilistic assumptions, it would be 1. That being the case, we set the constant C in our DBNS multiplication algorithm as 2. This is much smaller than the constants associated with the complexity analysis of the other subquadratic multiplication algorithms and points out that for medium-sized integers this algorithm has a clear potential to outperform all of the existing ones, including the shift-and-add algorithms.

The key point of the algorithm is the expression of one of the multiplicands in DBNS format with a restricted highest power of 3. The following example explains the encoding scheme using the DBNS tabular form.

Representation of 10,601 into DBNS format with the highest possible power of 3 and using two 7-bit blocks is shown in Figure 4.2 (note that we have used 1 to represent the nonzero digit, since a signed digit system will be used in

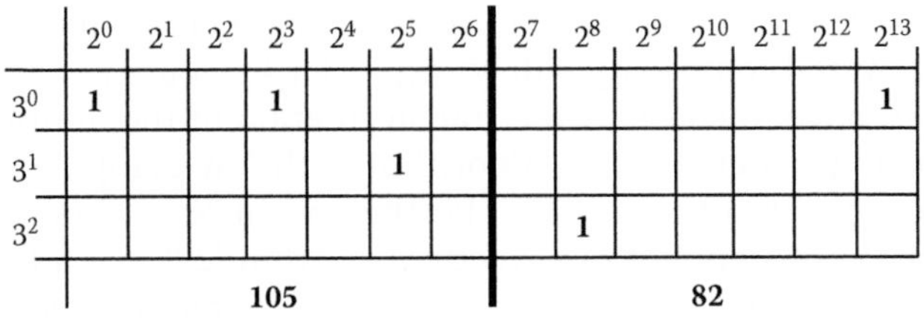

2^0 2^1 2^2 2^3 2^4 2^5 2^6 2^7 2^8 2^9 2^{10} 2^{11} 2^{12} 2^{13}
3^0
3^1
3^2
1
1
1
1
1
105
82

FIGURE 4.2
A DBNS representation of 10,601 using the blocking algorithm. (From V.S. Dimitrov, K.U. Jarvinen, J. Adikari, Area-Efficient Multipliers Based on Multiple-Radix Representations, *IEEE Transactions on Computers*, 60(2), 189–201, 2011. Copyright. With permission.)

general). The DBNS representation of $10601_{10} = 10100101101001_2$. This can be blocked as $10601 = 82 \times 2^7 + 105$ by using two blocks of length 7 bits and highest power of 3 equal to 2.

Since the conversion of the constant into DBNS form can be done very quickly, it makes sense to investigate the use of the above algorithm for general purpose multiplication.

4.5 General Multiplier Structure

The better asymptotic complexity of the DBNS-based multiplication algorithm guarantees that, eventually, it will outperform the shift-and-add-based algorithms for a certain range of multiplicands. The biggest practical problems are: (1) when it will happen, and (2) how to apply the algorithm as efficiently as possible in hardware.

In summary, the idea is to substitute the multiplication of B (the second multiplicand) by A into a succession of multiplications by several very sparse binary numbers—the rows of the DBNS matrix—and multiplications by some small number of powers of 3.

There are several parameters of the algorithm that have to be selected with extreme care. The first one is the window size, w. The best plan is to use a lookup table (LUT) approach for the conversion, so from a hardware viewpoint, it is desirable to use LUTs that are not too large. In principle, the larger the size of the LUT, the fewer additions/subtractions we will use. However, computational experiments show that when the LUT size grows to a certain point, the number of additions/subtractions does not decrease significantly, while the area complexity of the LUT grows exponentially.

VLSI complexity theory [39] suggests that one should try to minimize the AT^2 complexity measure, where A is the area complexity of the design and T is the time complexity. Current VLSI designs usually emphasize issues such as reduction of the power consumption, reduction of the critical path, etc., rather than AT^2. As we have pointed out in the introduction to this chapter, the time complexity of the multiplication problem is still unsolved. In the case of area-time complexity, the situation is even more unclear; however, there are some results that provide certain nontrivial lower bounds; below we refer to one of those results [40]:

Theorem 4.5: For any VLSI circuit that computes the middle bit of the multiplication of two *n*-bit integers, the area-time complexity satisfies the following lower bound: $AT^2 = \Omega(n^2)$.

Next, we give a general description of the proposed multiplier. Let A and B be two *k*-bit unsigned integers, i.e., A, B ∈ $[0, 2^{k-1}]$, and let C denote the 2*k*-bit result of the multiplication C = A·B. All proposed multipliers compute the entire product in parallel combinatorially, i.e., without registers or feedback loops. Although we consider only fully parallel multipliers in this report, designing digit-serial multipliers based on the new ideas would be straightforward.

Perhaps, the biggest advantage of the DBNS multiplier is the fact that it has a provably subquadratic complexity. Therefore it suffices to implement the algorithm for relatively small numbers and then approximate the break-even point with the reference multipliers by using the known complexity estimates for the two algorithms: binary based and DBNS based.

An alternative way to write the double-base number representation, obtained by the blocking algorithm, is given by the following formula:

$$A = \sum_{i=0}^{m} 3^i \left(\sum_{j=1}^{c(i)} s_i^{(j)} 2^{b_i^{(j)}} \right) \tag{4.2}$$

where m is the maximal power of 3 used; $c(i)$ is the number of binary exponents that has to be multiplied by 3^i, $0 \le i \le m$; $b_i^{(j)}$ is the *j*th binary exponent that corresponds to 3^i, $1 \le j \le c(i)$; and $s_i^{(j)} = \pm 1$ gives the sign of the term $2^{b_i^{(j)}}$.

Example 4.2

Use Equation (4.2) to represent 10601:

$m = 2$—the highest power of 3 used

$c(0) = 3$—the number of nonzero terms on the first row of the DBNS matrix (see Figure 4.2)

$b_0^{(1)} = 0, b_0^{(2)} = 3, b_0^{(3)} = 13$—the exponents of 2 that are used on the first row of the DBNS matrix

$s_0^{(1)} = 1, s_0^{(2)} = 1, s_0^{(3)} = 1$—the sign of the powers of 2 used on the first row of the DBNS matrix

$c(1) = 1$—the number of nonzero terms on the second row of the DBNS matrix

$b_1^{(1)} = 5$—the exponents of 2 that are used on the second row of the DBNS matrix

$s_1^{(1)} = 1$—the sign of the powers of 2 used on the second row of the DBNS matrix

$c(2) = 1$—the number of nonzero terms on the third row of the DBNS matrix

$b_2^{(1)} = 8$—the exponents of 2 that are used on the third row of the DBNS matrix

$s_2^{(1)} = 1$—the sign of the powers of 2 used on the third row of the DBNS matrix

The fact that different rows may (and usually do) require different numbers of terms complicates the design and contributes to much of the time and, especially, the area complexity. So, certain modifications of the encoding scheme have to be made to simplify the design, without affecting the main advantage: the asymptotic superiority over binary-based multiplication.

Let us generalize the above formula and consider the representation of A as in Equation (4.3):

$$A = \sum_{i=0}^{m} d^i \left(\sum_{j=1}^{c(i)} s_i^{(j)} 2^{b_i^{(j)}} \right) \tag{4.3}$$

That is, we apply the same idea; however, we multiply the elements of every row by suitably chosen integers, d_i. By "suitably chosen" we mean numbers that are specifically selected so as to optimize the performance of the multiplier for a particular window size, w. It is clear that if $d_i = 3^i$, then we have the representation outlined in Equation (4.2).

Let us clarify the importance of this new degree of freedom associated with the considerably unrestricted choice of the digits. If we, say, would like to use a window of size 7, then we have to make sure that any integer between 0 and 127 can be represented by using the corresponding number representation (Equation (4.2) or (4.3)). But if we want to use the double-base number representation in Equation (4.2), then we will have to use three terms, because certain numbers less than 127 cannot be represented as the sum or difference of two {2, 3}-integers (as we have pointed out above, the smallest positive integer with this property is 103). On the other hand, if one uses representations of the form of Equation (4.3), then it is sufficient to use digits $d_i = \{1, 3, 5, 7\}$ that guarantee a representation for every 7-bit integer of the form of Equation (4.3) by using at most two terms. The following fact will be used in the design, so we shall specifically acknowledge it:

Fact: Every nonnegative 7-bit integer can be represented in the form $z_1 \pm z_2$, where $z_1, z_2 \in \{1{\cdot}2^k, 3{\cdot}2^k, 5{\cdot}2^k, 7{\cdot}2^k, k = \{0, 1, \ldots, 7\}$.

Note: The smallest number for which the above fact is not valid is 137, i.e., an 8-bit number.

So, from a point of view of integer representations, this new number representation is more attractive compared to DBNS. In order to cover the same range (7-bit numbers) with the DBNS one must use the digit set {1, 3, 9, 27, 81} as shown in Table 2.1 (Chapter 2).

The encoders are the heart of the multiplier. The selection of the type of encoding to be used determines the construction of the remainder of the multiplier. In the following, we discuss the effects and trade-offs of parameter selection. If we use a window size 7, then one can encode every 7-bit number as in Table 4.1. The entire encoding table is given in Table 4.2, and it is used for coders of the multiplier in Figure 4.3. This encoding scheme deserves a few comments. First, it is very compact; it represents every 7-bit number by an 11-bit block of well-structured data. If one uses signed digit representations, then one has to substitute the 7-bit numbers with a 14-bit succession of zeros and ones, due to the necessity to use 2 bits per digit in the binary number system with digits 0, 1, and –1. Second, only one sign bit is required, because the first number, z_1, is positive by default. Third, for every 7-bit number we use exactly two terms, which greatly reduces the number of multiplexers and does not negatively affect the length of the critical path. The following example shows how to interpret Table 4.2.

Example 4.3

Encode the number 89 using Table 4.2.

From Table 4.2 we see that 89 is encoded in five blocks of bits as 01 101 11 0 000. The first two blocks encode the first term, z_1, and the remaining three blocks encode the second term, z_2. The first block, 01, shows that we select 3 (the second digit from the set {1, 3, 5, 7}), and the second block, 101, gives the power of two: $2^{101_2} = 32$. As a result, the first term becomes $3 \times 32 = 96$. The third block, 11, shows that we select 7 (the fourth digit in {1, 3, 5, 7}). The fourth block, 0, gives the sign of the second term, which, in this case, is negative. The fifth block, 000, gives the power of 2: $2^{000_2} = 1$. Hence, the second term is $-7 \times 1 = -7$. Therefore, 89 is represented as $96 - 7 = (3 \times 32) - (7 \times 1)$.

If one prefers to work with a matrix representation of numbers, then the following DBNS-like representation, given in Table 4.3, is useful. In this representation we use exactly one term from the first four rows and exactly one term from the following four rows.

Now we are in position to generalize this representation to (1) more blocks and (2) more flexible selection of digits. The reason why such a generalization is necessary is the following. It can be considered somewhat lucky that with this selection of digits {1, 3, 5, 7} we succeeded in covering the complete 7-bit dynamic range. As we have mentioned, 137 cannot be represented as

TABLE 4.2

The Complete 7-Bit Lookup Table

0	00 000 00 0 000	1	00 001 00 0 000	2	00 000 00 1 000	3	00 000 00 1 001
4	00 001 00 1 001	5	00 000 00 1 010	6	00 001 00 1 010	7	00 011 00 0 000
8	00 010 00 1 010	9	00 000 00 1 011	10	00 001 00 1 011	11	00 011 01 1 000
12	00 010 00 1 011	13	00 000 01 1 010	14	00 100 00 0 001	15	00 100 00 0 000
16	00 011 00 1 011	17	00 000 00 1 100	18	00 001 00 1 100	19	00 100 01 1 000
20	00 010 00 1 100	21	00 000 10 1 010	22	00 100 01 1 001	23	00 100 11 1 000
24	00 011 00 1 100	25	00 000 01 1 011	26	00 001 01 1 011	27	00 101 10 0 000
28	00 101 00 0 010	29	00 101 01 0 000	30	00 101 00 0 001	31	00 101 00 0 000
32	00 100 00 1 100	33	00 000 00 1 101	34	00 001 00 1 101	35	00 101 01 1 000
36	00 010 00 1 101	37	00 101 10 1 000	38	00 101 01 1 001	39	00 101 11 1 000
40	00 011 00 1 101	41	00 000 10 1 011	42	00 001 10 1 011	43	01 000 10 1 011
44	00 101 01 1 010	45	01 100 01 0 000	46	00 101 11 1 001	47	01 100 00 0 000
48	00 100 00 1 101	49	00 000 01 1 100	50	00 001 01 1 100	51	01 000 01 1 100
52	00 010 01 1 100	53	01 100 10 1 000	52	00 110 10 0 001	53	01 100 11 1 000
56	00 110 00 0 011	57	00 000 11 1 011	58	00 110 01 0 001	59	00 110 10 0 000
60	00 110 00 0 010	61	00 110 01 0 000	62	00 110 00 0 001	63	00 110 00 0 000
64	00 101 00 1 101	65	00 000 00 1 110	66	00 001 00 1 110	67	00 110 01 1 000
68	00 010 00 1 110	69	00 110 10 1 000	70	00 110 01 1 001	71	00 110 11 1 000
72	00 011 00 1 110	73	10 100 11 0 000	74	00 110 10 1 001	75	10 100 10 0 000
76	00 110 01 1 010	77	10 100 01 0 000	78	00 110 11 1 001	79	10 100 00 0 000
80	00 100 00 1 110	81	00 000 10 1 100	82	00 001 10 1 100	83	01 000 10 1 100
84	00 010 10 1 100	85	10 000 10 1 100	86	01 001 10 1 100	87	10 100 11 1 000
88	00 110 01 1 011	89	01 101 11 0 000	90	01 101 01 0 001	91	01 101 10 0 000
92	00 110 11 1 010	93	01 101 01 0 000	94	01 101 00 0 001	95	01 101 00 0 000
96	00 101 00 1 110	97	00 000 01 1 101	98	00 001 01 1 101	99	01 000 01 1 101
100	00 010 01 1 101	101	01 101 10 1 000	102	01 001 01 1 101	103	01 101 11 1 000
104	00 011 01 1 101	105	11 100 11 0 000	106	01 101 10 1 001	107	11 100 10 0 000
108	00 111 10 0 010	109	11 100 01 0 000	110	01 101 11 1 001	111	11 100 00 0 000
112	00 111 00 0 100	113	00 000 11 1 100	114	00 001 11 1 100	115	01 000 11 1 100
116	00 111 01 0 010	117	10 000 11 1 100	118	00 111 10 0 001	119	11 000 11 1 100
120	00 111 00 0 011	121	00 111 11 0 000	122	00 111 01 0 001	123	00 111 10 0 000
124	00 111 00 0 010	125	00 111 01 0 000	126	00 111 00 0 001	127	00 111 00 0 000

Source: V.S. Dimitrov, K.U. Jarvinen, J. Adikari, Area-Efficient Multipliers Based on Multiple-Radix Representations, *IEEE Transactions on Computers*, 60(2), 189–201, 2011. Copyright © 2011. With permission.

the sum or difference of two numbers belonging to the set $\{1{\cdot}2^k, 3{\cdot}2^k, 5{\cdot}2^k\}$ with integer $k \geq 0$; therefore we cannot straightforwardly extend this encoding to the 8-bit dynamic range. One will need either more summands or more digits, or both.

After examining many options, we have concluded that for multipliers it is optimal to have two summands (as in the above explained case with 7-bit

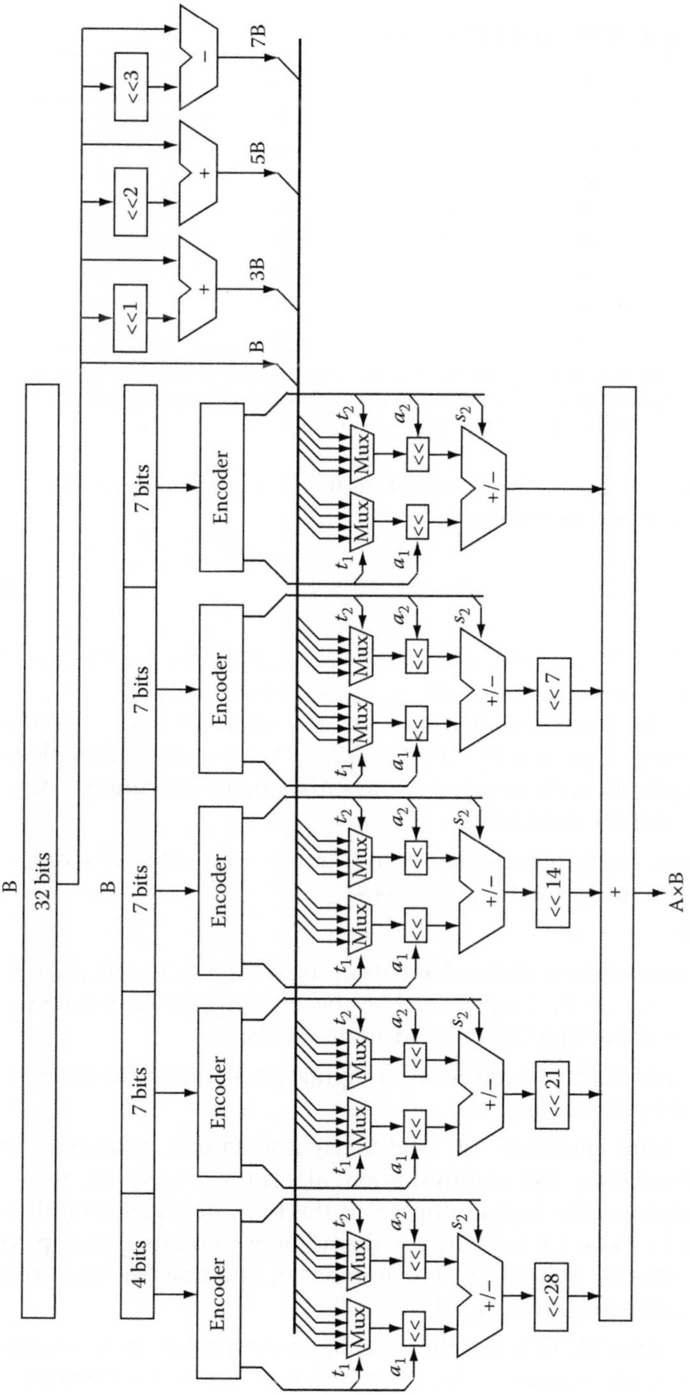

FIGURE 4.3
32 × 32-bit multiplier. (From V.S. Dimitrov, K.U. Jarvinen, J. Adikari, Area-Efficient Multipliers Based on Multiple-Radix Representations, *IEEE Transactions on Computers*, 60(2), 189–201, 2011. Copyright © 2011. With permission.)

TABLE 4.3

Representation of 89 with 7-Bit Encoding

	2^0	2^1	2^2–2^4	2^5
1	0	0	0	0
3	0	0	0	1
5	0	0	0	0
7	0	0	0	0
1	0	0	0	0
3	0	0	0	0
5	0	0	0	0
7	–1	0	0	0

Source: V.S. Dimitrov, K.U. Jarvinen, J. Adikari, Area-Efficient Multipliers Based on Multiple-Radix Representations, *IEEE Transactions on Computers*, 60(2), 189–201, 2011. Copyright © 2011. With permission.

numbers) and a carefully selected set of digits. This particular encoding can be formally expressed as follows:

$$A = \pm z_1 \pm z_2 \tag{4.4}$$

where $z_1 = \{a_1 \cdot 2^k, a_2 \cdot 2^k, \ldots, a_s \cdot 2^k\}$, $z_2 = \{b_1 \cdot 2^k, b_2 \cdot 2^k, \ldots, b_l \cdot 2^k\}$ for $k = 0,1\ldots, w$. The determination of the sets of $D_a = \{a_1, a_2, \ldots, a_s\}$ and $D_b = \{b_1, b_2, \ldots, b_1\}$ is the cornerstone of the proposed algorithm. In the above example, with 7-bit integers, the sets of digits are $D_a = D_b = \{1, 3, 5, 7\}$. For a successful implementation of multiplication, these sets have to satisfy many conditions. We will provide a few of these conditions:

a. Mandatory: Every number between 0 and $2^w - 1$ must have a representation in the form of Equation (4.4).
b. Optional:
    1. The computation of the elements of the set $U = \{a_1 \cdot B. a_2 \cdot B, \ldots, a_s \cdot B, b_1 \cdot B, b_2 \cdot B, \ldots, b_l \cdot B\}$ should be achievable by using the minimal possible number of additions and subtractions.
    2. The critical path of the above computation should be as small as possible, ideally one.
    3. The total number of additions/subtractions required in implementing the multiplication algorithm based on representation of the first multiplier in the form of Equation (4.4) is $2\lceil k/w \rceil + s + l + t + 1$, where t is the number of digits that appear in both sets, D_a and D_b. The number of additions/subtractions should also be minimized.
    4. If it is possible to select those digits in such a way as to encode every w-bit number in the form z (that is, the first summand is

always positive), then it will lead to a smaller area complexity of the design.

5. If it is possible to fix the signs of both the summands, z_1 and z_2 (that is, if we represent every w-bit integer as either $z_1 + z_2$ or $z_1 - z_2$), then we can expect further hardware simplifications due to the elimination of the necessity to process the sign of the second summand.

To see how delicate the digit selection procedure is, we will consider the case of 8-bit windows applied to 64×64-bit multiplication. To ensure that condition b.1 is satisfied, it is clear that one should choose as many identical digits as possible. However, if we want all the digits to be identical, that is, $s = l$ and $a_1 = b_1$ for $i = 1, 2, \ldots, s$, then the computational experiments show that one will need at least seven digits to guarantee a representation for every 8-bit number—the mandatory condition a. In this case one will need at most 22 additions/subtractions to implement the 64×64-bit multiplication. Not that this estimate of 22 is bad, but it is not optimal.

The solution provided below, found via an extensive combinatorial search (more than 10 billion possible digit combinations have been analyzed), shows one nontrivial example of digit selection.

Fact: Every 8-bit nonnegative integer can be represented with the form $z_1 \pm z_2$, where $z_1 \in \{1{\cdot}2^k, 3{\cdot}2^k, 5{\cdot}2^k, 7{\cdot}2^k, 11{\cdot}2^k, 13{\cdot}2^k\}$, $\{1{\cdot}2^k, 3{\cdot}2^k, 5{\cdot}2^k, 7{\cdot}2^k, 11{\cdot}2^k, 119{\cdot}2^k\}$ for $k = \{0, 1, \ldots. 8\}$

Note that the first summand, z_1, is always positive, so we do not need to assign a special bit for its sign. The two sets of digits are $D_a = \{1, 3, 5, 7, 11, 13\}$ and $D_b = \{1, 3, 5, 7, 11, 119\}$. The B processing can be carried out as follows (where (B << s) means B left shifted s times or $B.2^s$):

$$3 \cdot B = (B = 1) + B = B_1 \qquad /1 \text{ addition}$$

$$5 \cdot B = (B = 2) + B \qquad /1 \text{ addition}$$

$$7 \cdot B = (B = 3) - B = B_2 \qquad /1 \text{ subtraction}$$

$$11 \cdot B = (B = 2) + B_2 \qquad /1 \text{ addition}$$

$$13 \cdot B = (B_1 = 2) + B \qquad /1 \text{ addition}$$

$$119 \cdot B = (B_2 = 4) + B_2 \qquad /1 \text{ addition}$$

TABLE 4.4

Representations with the Smallest Maximum Numbers of Additions/Subtractions for $k = 64$, 128, and 256

k	w	Digit Sets	Additions/Subtractions
64	11	{1,3,5,7}, {1,3,5,7}, {1,3,5,7}	20
128	8	{1,3,5,7,11,13}, {1,3,5,7,11,119}	37
256	8	{1,3,5,7,11,13}, {1,3,5,7,11,119}	69

Source: V.S. Dimitrov, K.U. Jarvinen, J. Adikari, Area-Efficient Multipliers Based on Multiple-Radix Representations, *IEEE Transactions on Computers*, 60(2), 189–201, 2011. Copyright © 2011. With permission.

The overall number of additions/subtractions is six, which is minimal. The critical path is two, which is also minimal among all digit sets that ensure the representation of every 8-bit integer and that require only six additions to generate all the multiples of 5. Now, it is easy to estimate that this encoding scheme guarantees the implementation of multiplication by any 64-bit number with at most 21 additions/subtractions, and what is even more important is that it can be done very quickly.

In the case of the 64 × 64-bit multiplication, this new encoding scheme does not have advantages over the one proposed with the 7-bit windowing, which also guarantees 21 additions/subtractions with a less complicated encoding scheme and a smaller LUT. However, in the case of the 128 × 128-bit multiplication, the new encoding is superior; it requires only 37 additions/subtractions, whereas the 7-bit windowing requires 40 additions/subtractions.

Based on the above theory, and many experiments that we have implemented, we would recommend the use of windows of size 6 (at least) and 11 (at most) for multiplication of integers of medium size: 64 × 64, 128 × 128, and 256 × 256. For sizes that would be useful, e.g., in RSA cryptography, we may need larger windows if the hardware resources allow for that.

Our results allow us to provide some considerably nontrivial upper bounds on the number of additions/subtractions sufficient to implement a $k \times k$-bit multiplication for different values of k. Table 4.4 provides information for $k = 64$, 128, and 256 and the corresponding encoding of the multiplicands that guarantee these upper bounds.

4.6 Results and Comparisons

Several multipliers using the above representations were described in VHDL in order to find out how they perform in practice. The representations used in these multipliers are collected in Table 4.5. They were carefully selected from many possibilities because they appeared to have very attractive features in theory or practice, as discussed in the previous sections. Notice that the design depicted in Figure 4.3 is a 32-bit version of mult_7b2d.

TABLE 4.5
Selected Representations

Name	w	Digit Sets
mult_6b2d7	6	{1,3,5,7}, {1,3}[a]
mult_6b2d9	6	{1,3,5,9}, {1,3}
mult_6bsms	6	{1,3,5,7}, {1,3,5,7}[a,b]
mnlt_7b2d	7	{1,3,5,7}, {1,3,5,7}[a]
mult_7bsms	7	{1,3,5,7,89,125}, {1,3,5,7,89,125}[a,b]
mult_8b2dd	8	{1,3,5,7,11,13}, {1,3,5,7,11,119}[a]
mult_8b2di	8	{1,3,5,7,11,15,19,25}, {1,3,5,7,11,15,19,25}[a]
mult_8b3d	8	{1}, {3}, {7,17}
mult_8bsms	8	{1,3,5,7,11,13,15}, {1,3,5,7,11,13,15}[a,b]
mult_9b2d	9	{1,3,5,7,11,13,15}, {1,3,5,7,11,13,15}[a]
mult_11b3d	11	{1,3,5,7}, {1,3,5,7}, {1,3,5,7}[a]

Source: V.S. Dimitrov, K.U. Jarvinen, J. Adikari, Area-Efficient Multipliers Based on Multiple-Radix Representations, *IEEE Transactions on Computers*, 60(2), 189–201, 2011. Copyright © 2011. With permission.

[a] The first term is always positive.

[b] The second term is always negative.

32-bit and 64-bit multipliers based on the representations of Table 4.6 were synthesized for both 0.18 μm complementary metal oxide semiconductor (CMOS) and Altera Cyclone III Field-Programmable Gate Array (FPGA). The proposed algorithms are mainly aimed at ASIC design; however, FPGA designs are faster to evaluate and relatively inexpensive, and at the same time, they serve as a guiding line for possible optimizations and improvements of the algorithms.

Typically, the presented multipliers either have smaller operands (e.g., 8 bit or 16 bit), or they are synthesized for a different technology. Hence, we synthesized our own reference multipliers in order to get a clearer picture of how our multipliers compare with the existing solutions. These reference multipliers were realized by synthesizing the VHDL statement: $c <= a * b$. Because multiplication is such a frequently used operation, it is undoubtedly well optimized by all logic synthesis programs. Hence, the reference multipliers qualify as valid comparison points for our multipliers.

4.7 Some Multiplier Designs

4.7.1 180 nm CMOS Technology

Area complexity, critical path delays, and power consumptions of our 32-bit and 64-bit multipliers are collected in Table 4.6. These results were produced

TABLE 4.6

Results in 0.18/μm CMOS

	32 × 32 Bit				64 × 64 Bit			
	Area		Power		Area		Power	
Design	μm²	Ratio	mW	Ratio	μm²	Ratio	mW	Ratio
mult_ref	163,133	1.00	6.33	1.00	989,039	1.00	44.87	1.00
mult_6b2d7	110,533	0.68	4.05	0.64	433,942	0.44	20.76	0.46
mult_6b2d9	128,735	0.79	4.70	0.74	486,270	0.49	22.65	0.50
mult_6bsms	104,432	0.64	4.05	0.64	418,348	0.42	21.44	0.48
mult_7b2d	114,458	0.70	4.24	0.67	407,082	0.41	19.53	0.44
mult_7bsms	132,194	0.81	4.81	0.76	494,702	0.50	23.68	0.53
mult_8b2dd	139,183	0.85	4.77	0.75	489,876	0.50	22.89	0.51
mult_8b2di	146,139	0.90	4.99	0.79	480,289	0.49	22.71	0.51
mult_8b3d	156,574	0.96	5.45	0.86	523,985	0.53	23.12	0.52
mult_8bsms	136,206	0.83	4.84	0.76	473,107	0.48	22.43	0.50
mult_9b2d	146,162	0.90	4.82	0.76	488,489	0.49	21.88	0.49
mult_11b3d	218,717	1.34	5.46	0.86	631,174	0.64	21.49	0.48

Source: V.S. Dimitrov, K.U. Jarvinen, J. Adikari, Area-Efficient Multipliers Based on Multiple-Radix Representations, *IEEE Transactions on Computers*, 60(2), 189–201, 2011. Copyright © 2011. With permission.

by synthesizing VHDL with 0.18 μm CMOS libraries with the target clock frequency set to 50 MHz.

When compared to the reference multipliers, the results in Table 4.6 demonstrate clear advantages in both area and delay. The best encodings are mult_6sms for the 32-bit multiplication and mult_7b2d for the 64-bit multiplication.

The results show the delicacy involved in selecting the representations. The quality of the results varies considerably even between representations, which, at first sight, have only small differences. The effects of the conditions discussed in Section 4.5 are clearly visible in the results. For instance, the "something minus something" (sms) encodings, where the first term is always positive and the second term is negative, show a clear advantage over other encodings with the same w. The only exception is $w = 7$ for which there exists a particularly elegant encoding with small digit sets (mult_7b2d); in this case, the smaller digit sets weigh more than the fixed sign. The sizes of the encoders start to play a significant role in the area complexity when w increases. This diminishes the feasibility of representations with large w, such as mult_11b3d, although they appear attractive in theory because of the low total number of additions/subtractions.

The 64-bit reference multiplier could not meet the clock constraint of 50 MHz. The synthesizer had to put in extra effort when trying to achieve this constraint. As a consequence, the area of the reference multiplier is approximately six times larger for the 64-bit multiplier than for the 32-bit multiplier.

This is more than one would expect according to theory, which predicts a fourfold increase when the width doubles. If the increase followed the theory, the area of the 64-bit multiplier would be approximately 650,000 μm^2 and the best of our 64-bit multipliers (mult_7b2d) would have an area of 63% of the reference. Hence, the advantages would be clear even in that case.

4.7.2 FPGA Implementation

We present results here for implementation of our multipliers using FPGAs for design iterations and prototyping. The results on FPGAs were not as spectacular as the above results, but they deserve attention nonetheless. The results for our best multiplier and the reference multiplier on the Altera Cyclone III EP3C40F780C7 are collected in Table 4.7. They were generated by compiling the designs in Quartus II 8.1.

Contrary to the results presented above, we were unable to better the performance of the reference multipliers when implemented on our target FPGA. There are several possible reasons for this. First, our multiplier architecture is rather ill-suited for FPGAs where multiplexers are notoriously area demanding. Second, the synthesis is well optimized to map traditional multipliers to the FPGA fabric efficiently, whereas it maps our multipliers as arbitrary logic; as a consequence, our multipliers use the FPGA less efficiently.

Large windows, w, are particularly unsuitable for FPGAs because the sizes of the encoders begin to dominate; the best results were achieved with mult_6bsms which uses window size $w = 6$. Table 4.7 shows the results for both mult_ref and mult_6bsms.

Table 4.7 shows that the gap between the reference multiplier and our design is diminishing. The reference multipliers appear to adhere strongly to the heuristic rule that doubling the size of the operands leads to quadrupling the area complexity: 5706/1423 = 4.0098. For our multiplier, the area complexity increment is 8580/2302 = 3.727. Encodings with larger windows have even smaller increment factors, although the absolute area consumptions are larger. In terms of time complexity, again, the gap is diminishing:

TABLE 4.7

Results on Altera Cyclone III FPGA (only the best one, mult_6bsms, is included)

	32 × 32 Bit				64 × 64 Bit			
	Area		Delay/Power		Area		Delay/Power	
Design	LUTs	Ratio	ns	Ratio	LUTs	Ratio	ns	Ratio
mult_ref	1,423	1.00	18.39	1.00	5,706	1.00	2,617	1.00
mult_6bsms	2,302	1.62	21.03	1.14	8,580	1.50	27	1.04

Source: V.S. Dimitrov, K.U. Jarvinen, J. Adikari, Area-Efficient Multipliers Based on Multiple-Radix Representations, *IEEE Transactions on Computers*, 60(2), 189–201, 2011. Copyright © 2011. With permission.

our multiplier is clearly behind in the case of the 32-bit multiplication but almost level at 64 bits.

Experiments with different bit widths allow us to predict that our algorithm would outperform algorithms based on classical binary arithmetic for the 1024 × 1024-bit multiplication in terms of area complexity; the break-even point for delay is probably much smaller. The only reason why we cannot evaluate this in practice is that such a design would be too large to fit into any existing FPGA. In the very near future this situation will undoubtedly change, letting us verify this assumption. However, even in that case, our multipliers will likely remain impractical on FPGAs because other solutions (such as digit-serial multipliers or embedded hardwired multipliers) most probably offer more competitive results.

4.8 Example Applications

Possible applications of our new multiplier are basically unlimited, because multiplication is such a central operation in most digital systems. However, the fact that our multipliers become better than the existing solutions only when the operands are wider than a certain threshold naturally sets some limits for the applications of our new multipliers. We demonstrated above that our multiplier is already superior if the operands are 32 bits or wider, so the threshold is obviously less than 32 bits. The exact value of the threshold depends on many things, such as the technology used in implementation, and determining exact thresholds will be a topic for future research. Nevertheless, the results presented above already demonstrate that the threshold is small enough to cover numerous applications of today's systems, including microprocessors, digital signal processing systems and processors, cryptography, etc.

One particularly interesting application is the possibility to use our multiplier in floating point operations. The floating point number systems used in practice typically represent numbers as described in the IEEE Standard for Floating-Point Arithmetic (IEEE 754) [41], which includes 32-bit (single precision), 64-bit (double precision), and 128-bit (quadraple precision) versions. In all of them, one bit signifies the sign. The exponent is represented with 8, 11, or 15 bits and the fraction is given by 23, 52, or 112 bits for single, double, and quadruple precision, respectively [41]. A floating point multiplication requires a multiplication of the fractions, e.g., a 52 × 52-bit multiplication for double precision, and consequently, a floating point processor must have support for multiplications with large operands. Clearly, the widths used, at least in the double and quadruple precision formats, exceed the threshold, when our multipliers become superior, and could therefore benefit from the results presented in this chapter.

4.9 Summary

In this chapter we have introduced one very promising application of the double-base number system, namely, efficient computation of the product of two integers. The main theoretical advantage of the DBNS—its sparseness—plays a critical role in the design of the algorithms and proving their efficacy. The analysis of the DBNS multiplication algorithm rigorously shows that it is asymptotically faster than the algorithms based on the shift-and-add approach. What is much more important is that the asymptotic advantage is not associated with a large increase of the constant associated with the big-O notation. The break-even point where the new approach starts to outperform the available method is, of course, platform dependent. For dynamic ranges such as 64×64, 128×128, 192×192 (which is particularly useful in crypto applications), and higher, the DBNS-based multiplication leads to considerable improvements. There are many possibilities for additional modifications that can lead to even better estimates. For example, one can try to use Booth encoding of the multiplicand and then convert it into a DBNS format. It is also possible to combine the algorithm proposed with Karatsuba-like decomposition of the product obtained. All these are wonderful opportunities for further investigations.

References

1. A.N. Kolmogorov, Asymptotic Characteristics of Some Completely Bounded Metric Spaces, *Doklady Akademii Nauk. SSSR*, 108, 585–589, 1956.
2. A. Karatsuba, Y. Ofman, Multiplication of Multidigit Numbers on Automata, *Soviet Physics Dokiady*, 7(7), 595–596, 1963.
3. D.E. Knuth, *Art of Computer Programming: Seminumerical Algorithms*, 3rd ed., vol. 2, Addison-Wesley Professional, Reading, MA, 1997.
4. D. Zuras, More on Squaring and Multiplying Large Integers, *IEEE Transactions on Computers*, 43(8), 899–908, 1994.
5. V. Strassen, Gaussian Elimination Is Not Optimal, *Numerische Mathematik*, 13(3), 354–356, 1969.
6. A. Schönhage, V. Strassen, Schnelle Multiplikation Großer Zahlen, *Computing*, 7, 281–292, 1971.
7. M. Fürer, Faster Integer Multiplication, in *Proceedings of the 39th Annual ACM Symposium on Theory of Computing*, San Diego, 2007, pp. 57–66.
8. S. A. Cook, On the Minimal Computation Time of Functions, PhD dissertation, Harvard University, 1966.
9. V. Lefévre, *Multiplication by an Integer Constant*, INRIA, a CCSD electronic archive server based on Laboratorie d'Lnformatique du Parellelisme, Lyon, France, Technical Report, 2000.

10. R. Pinch, Asymptotic Upper Bound for Multiplier Design, *Electronics Letters,* 32(5), 420–421, 1996.
11. V.S. Dimitrov, L. Imbert, A. Zakaluzny, Multiplication by a Constant Is Sublinear, in *Proceedings of the 18th IEEE Symposium on Computer Arithmetic,* ARITH18, June 2007, pp. 261–268.
12. V.S. Dimitrov, K.U. Jarvinen, J. Adikari, Area-Efficient Multipliers Based on Multiple-Radix Representations, *IEEE Transactions on Computers,* 60(2), 189–201, 2011.
13. V.S. Dimitrov, L. Imbert, P.K. Mishra, The Double-Base Number System and Its Application to Elliptic Curve Cryptography, *Mathematics of Computation,* 77(262), 1075–1104, 2007.
14. V.S. Dimitrov, G.A. Jullien, and W.C. Miller, An Algorithm for Modular Exponentiation, *Information Processing Letters,* 66(3), 155–159, 1998.
15. A. Baker, Linear Forms in the Logarithms of Algebraic Numbers IV, *Mathematika,* 16, 204–216, 1968.
16. V.S. Dimitrov, T.V. Cooklev, Hybrid Algorithm for the Computation of the Matrix Polynomial $I + A + \cdots + A^{N-1}$, *IEEE Transactions on Circuits and Systems I: Fundamental Theory and Applications,* 42(7), 377–380, 1995.
17. V.S. Dimitrov, G.A. Jullien, Loading the Bases: A New Number Representation with Applications, *Circuits and Systems Magazine,* 3(2), 6–23, 2003.
18. G. Jullien, V.S. Dimitrov, B. Li, W.C. Miller, A. Lee, and M. Ahmadi, A Hybrid DBNS Processor for DSP Computation, in *Proceedings of the IEEE International Symposium on Circuits and Systems,* ISCAS '99, July 1999, vol. 1, pp. 5–8.
19. R. Muscedere, V.S. Dimitrov, G.A. Jullien, W.C. Miller, Efficient Techniques for Binary-to-Multidigit Multidimensional Logarithmic Number System Conversion Using Range-Addressable Look-Up Tables, *IEEE Transactions on Computers,* 54(3), 257–271, 2005.
20. R. Muscedere, V.S. Dimitrov, G.A. Jullien, W.C. Miller, A Low-Power Two-Digit Multi-Dimensional Logarithmic Number System Filterbank Architecture for a Digital Hearing Aid, *EURASIP Journal on Applied Signal Processing,* 18, 3015–3025, 2005.
21. R. Muscedere, V.S. Dimitrov, G.A. Jullien, W.C. Miller, M. Ahmadi, On Efficient Techniques for Difficult Operations in One and Two-Digit DBNS Index Calculus, in *Proceedings of the 34th Asilomar Conference on Signals, Systems and Computers,* 2000, vol. 2, pp. 870–874.
22. S.M. Kilambi, B. Nowrouzian, A Novel Genetic Algorithm for Optimization of FRM Digital Filters over DBNS Multiplier Coefficient Space Based on Correlative Roulette Selection, in *Proceedings of the IEEE International Symposium on Signal Processing and Information Technology,* August 2006, pp. 228–231.
23. P. Mercier, S.M. Kilambi, B. Nowrouzian, Optimization of FRM FIR Digital Filters over CSD and CDBNS Multiplier Coefficient Spaces Employing a Novel Genetic Algorithm, *Journal of Computers,* 2(7), 20–31, 2007.
24. V.S. Dimitrov, K.U. Järvinen, M.J. Jacobson Jr., W.F. Chan, Z. Huang, Provably Sublinear Point Multiplication on Koblitz Curves and Its Hardware Implementation, *IEEE Transactions on Computers,* 57(11), 1469–1481, 2008.
25. V.S. Dimitrov, L. Imbert, P.K. Mishra, Efficient and Secure Elliptic Curve Point Multiplication Using Double-Base Chains, in *Proceedings of the Advances in Cryptology,* ASIA CRYPT '05, Series Lecture Notes in Computer Science, vol. 3788, Springer, Berlin, 2005, pp. 59–78.

26. R. Avanzi, V.S. Dimitrov, C. Doche, F. Sica, Extending Scalar Multiplication Using Double Bases, in *Proceedings of the Advances in Cryptology*, ASIA CRYPT '06, Series Lecture Notes in Computer Science, vol. 4284, Springer, Berlin, 2006, pp. 130–144.
27. M. Ciet, F. Sica, An Analysis of Double Base Number Systems and a Sublinear Scalar Multiplication Algorithm, in *Proceedings of the Progress in Cryptology*, Mycrypt '05, Series Lecture Notes in Computer Science, vol. 3715, Springer, Berlin, 2005, pp. 171–182.
28. C. Doche, L. Imbert, Extended Double-Base Number System with Applications to Elliptic Curve Cryptosystem, in *Proceedings of the Progress in Cryptology*, INDOCRYPT '06, Series Lecture Notes in Computer Science, vol. 4329, Springer, Berlin, 2006, pp. 335–348.
29. V. Berthé, L. Imbert, On Converting Numbers to the Double-Base Number System, *Proceedings of SPIE*, 5559, 70–78, 2004.
30. V.S. Dimitrov, T.V. Cooklev, Two Algorithms for Modular Exponentiation Using Nonstandard Arithmetic, *IEICE Transactions on Fundamentals*, E78-A, 82–87, 1995.
31. C.-Y. Chen, C.-C. Chang, W.-P. Yang, Hybrid Method for Modular Exponentiation with Precomputation, *Electronics Letters*, 32(6), 540–541, 1996.
32. C. Doche, D. Kohel, F. Sica, Double-Base Number System for Multi-Scalar Multiplications, in *Proceedings of the EuroCrypt '09*, Series Lecture Notes in Computer Science, vol. 5479, Springer, Berlin, 2009, pp. 502–519.
33. C. Doche, L. Habsieger, A Tree-Based Approach for Computing Double-Base Chains, in *Proceedings of the 13th Australasian Conference on Information Security and Privacy*, ACISP '08, Springer-Verlag, Berlin, 2008, pp. 433–446.
34. J. Adikari, V.S. Dimitrov, L. Imbert, Hybrid Binary-Ternary Joint Form and Its Application in Elliptic Curve Cryptography, in *Proceedings of the 19th IEEE Symposium on Computer Arithmetic, 2009*, ARITH19, June 2009, pp. 76–83.
35. K.W. Wong, E.C.W. Lee, L. Cheng, X. Liao, Fast Elliptic Scalar Multiplication Using New Double Base Chain and Point Halving, *Applied Mathematics and Computation*, 183(2), 1000–1007, 2006.
36. C. Zhao, F. Zhang, J. Huang, Efficient Tate Pairing Computation Using Double-Base Chains, *Science in China Series F: Information Sciences*, 51(8), 1009–2757, 2008.
37. D.J. Bernstein, T. Lange, Analysis and Optimization of Elliptic-Curve Single-Scalar Multiplication, *Finite Fields and Applications, Contemporary Mathematics*, 461, 1–19, 2008.
38. R. Bernstein, Multiplication by Integer Constants, *Software: Practice and Experience*, 16(7), 641–652, 1986.
39. J.D. Ullman, *Computational Aspects of VLSI*, W.H. Freeman & Co., New York, 1984.
40. I. Wegener, R. Pruim, *Complexity Theory: Exploring the Limits of Efficient Algorithms*, Springer-Verlag, Secaucus, NJ, 2005.
41. IEEE Computer Society, Standard for Floating-Point Arithmetic, IEEE 754-2008.
42. V.S. Dimitrov, K.U. Jarvinen, J. Adikari, Area-Efficient Multipliers Based on Multiple-Radix Representations, *IEEE Transactions on Computers*, 60(2), 189–201, 2011.

5

The Multidimensional Logarithmic Number System (MDLNS)

5.1 Introduction

We now return to the two-dimensional index calculus that was introduced in Chapter 2. As a quick review, we started our discussion with the double-base index calculus where the number $x = \sum d_{i,j} 2^i 3^j$ is represented by the indices of the two bases for each nonzero digit. That is, $x = \sum_{i=1}^{n} 2^{a_i} 3^{b_i}$ for n nonzero digits, where each digit is represented by the tuple $x_i \rightarrow \{a_i, b_i\}$. In order to accommodate signed numbers, we include a sign bit, $s_i \in \{-1, 1\}$, so that each digit position in $x_i = \sum_{i=1}^{n} s_i 2^{a_i} 3^{b_i}$ is represented by the triple $s_i 2^{a_i} 3^{b_i} \rightarrow (s_i, a_i, b_i)$. We also provided Theorem 2.3, which shows that every real number can be approximated, to arbitrary precision, with just a single such triple. We are now ready to delve into the theory and application of the multidimensional logarithmic number system (MDLNS). Part of this chapter is based on reference [21] (© CRC Press 2006).

5.2 The Multidimensional Logarithmic Number System (MDLNS)

We start formally with some definitions.

Definition 5.1: An s-integer is a number whose largest prime factor does not exceed the sth prime number.

We now modify this definition to the representation of arbitrary real numbers:

Definition 5.2: Modified 2-integers are numbers of the form $2^a p^b$, p an odd integer, where a and b are signed integers.

The next definition offers the most general representation scheme we will consider in this chapter.

Definition 5.3: A representation of the real number x in the form

$$x = \sum_{i=1}^{n} s_i \prod_{j=1}^{B} p_j^{e_j^{(i)}} \tag{5.1}$$

where $s \in \{-1, 0, 1\}$ and $p_j, e_j^{(i)}$ are integers, is called a multidimensional n-digit logarithmic (MDLNS) representation of x. B is the number of bases used (at least two; the first one, that is, p_1, will always be assumed to be 2 for implementation efficiency).

The next two definitions are special cases of Definition 5.3; the representation schemes defined by them will be used extensively in this chapter.

Definition 5.4: An approximation of a real number, x, as a signed modified 2-integer $s2^a p^b$, is called a two-dimensional logarithmic representation of x.

Definition 5.5: An approximation of a real number, x, as a sum of signed modified 2-integers $\sum_{i=i}^{n} s_i 2^{a_i} p^{b_i}$, is called an n-digit two-dimensional logarithmic representation of x. $n = 2$ will be a special case.

It is important to note that an extension of the classical LNS to a multi-digit representation does not provide any inherent advantages in terms of complexity reduction. Arnold et al. were the first to consider a similar representation scheme [1] in the case of classical LNS (we shall call it a two-component LNS). Although it leads to some reduction of the dynamic range of the exponents (correspondingly, a reduction of the lookup table sizes for the difficult operations), the number of bits required by the larger exponent to store the integer number, x, is approximately $\log(x)$. The storage reduction in the two-component LNS (as opposed to the one-component LNS) comes from the observation that in the one-component LNS one needs approximately $\log_2(x) + \log_2(\log(x))$ bits to encode x.

We will demonstrate that hardware complexity for the MDLNS is exponentially dependent on the size of the nonbinary-base exponents; we clearly have a potential for quite a considerable hardware reduction providing that the dynamic range of the nonbinary exponents is reduced as much as possible.

5.3 Arithmetic Implementation in the MDLNS

To a great extent we follow the procedures developed for implementing arithmetic in the LNS over the several decades that the LNS has been studied [2–5].

To summarize, a 2DLNS representation provides a triple, $\{s_i, a_i, b_i\}$, for each digit, where s_i is the sign bit and a_i, b_i are the exponents of the binary and nonbinary bases, and a number, x, is approximated by Definition 5.5.

5.3.1 Multiplication and Division

MDLNS multiplication and division are the simplest of the arithmetic operations. The equations for multiplication and division, given a single-digit 2DLNS representation of $x = \{s_x, a_x, b_x\}$ and $y = \{s_y, a_y, b_y\}$, are [6]:

$$x \cdot y = \{s_x \cdot s_y, a_x + a_y, b_x + b_y\} \tag{5.2}$$

$$x \div y = \{s_x \cdot s_y, a_x - a_y, b_x - b_y\} \tag{5.3}$$

Equations (5.2) and (5.3) show that single-digit 2DLNS multiplication can be implemented in hardware using two independent binary adders and simple logic for the sign correction. As we start to add digits to the representation we will face the equivalent of implementing multiplication with the addition of partial products. A two-digit representation will produce four independent partial products that will have to be added, and since addition is an expensive operation in the MDLNS, we try to reduce the cost of this process as much as possible.

5.3.2 Addition and Subtraction

Unfortunately, as with the classical LNS, addition and subtraction operations are not as simple as multiplication and division operations. Traditionally, addition and subtraction must be handled through a set of identities and lookup tables (LUTs). The identities are [6]

$$\begin{aligned}(2^{a_1} \cdot p^{b_1}) + (2^{a_2} \cdot p^{b_2}) &= (2^{a_1} \cdot p^{b_1}) \cdot (1 + 2^{a_2 - a_1} \cdot p^{b_2 - b_1}) \\ &\approx (2^{a_1} \cdot p^{b_1}) \cdot \Phi(a_2 - a_1, b_2 - b_1)\end{aligned} \tag{5.4}$$

$$\begin{aligned}(2^{a_1} \cdot p^{b_1}) - (2^{a_2} \cdot p^{b_2}) &= (2^{a_1} \cdot p^{b_1}) \cdot (1 - 2^{a_2 - a_1} \cdot p^{b_2 - b_1}) \\ &\approx (2^{a_1} \cdot p^{b_1}) \cdot \Psi(a_2 - a_1, b_2 - b_1)\end{aligned} \tag{5.5}$$

The operators Φ and Ψ are LUTs that store the precomputed 2DLNS values of

$$\begin{aligned}\Phi(a,b) &= 1 + (2^a \cdot p^b) \\ \Psi(a,b) &= 1 - (2^a \cdot p^b)\end{aligned} \tag{5.6}$$

Note that the identities of Equations (5.5) and (5.6) reduce a binary (two-input) LUT to a unary LUT (the slide rule trick from Chapter 1); this reduces the size of the addition LUT by half of the address bits (for equal word length addends); even so, the tables can still be quite large.

The use of large LUTs, implemented through the use of ROMs, for the evaluation of addition and subtraction operations is the most straightforward approach in systems such as the LNS [1] (other techniques for evaluating the functions of Equation (5.6) can be found in [7] and [8]). The large ROM table technique is only feasible for very small ranges of 2DLNS numbers. It is more practical, in most cases, to convert the 2DLNS numbers to binary and perform the addition and subtraction using a binary representation.

The conversions from 2DLNS to binary will still require a LUT, but one that is much smaller than required for handling 2DLNS addition and subtraction. The LUT is used to convert the second base portion of the 2DLNS number into a binary representation. Therefore the size of the LUT is dependent on the number of bits used to represent the second base exponent.

It should be noted that a new architecture was proposed in [9] that significantly reduces the size of the tables required for 2DLNS addition and subtraction.

5.3.4 Approximations to Unity

A very fundamental difference between the classical LNS and multidimensional LNS is the possibility to find nontrivial approximations of unity in the MDLNS. These can be used to constrain the dynamic range of exponents during general computations, a feature that is essential for fixed-exponent dynamic range in DSP applications. This is equivalent to scaling in other number systems where a fixed word length is used to hold the result of arithmetic calculations. Rounding errors will be introduced when the calculations require a larger word length to hold the precise result.

From Chapter 2, Theorem 2.3, we know that unity can be approximated with arbitrary precision as a 2-integer. In fact, *both* bases can be changed and the theorem will still remain valid. Here we expand the discussion of these approximants within the MDLNS, and discuss some results.

As an example with 8-bit exponents, consider the generation of a sequence of successive values of possible one-digit MDLNS (2-integer) values. This is easy to do with a spreadsheet, although we do require the generation of 2^{16} rows! The algorithm is quite simple (see Figure 5.1), though the results are very interesting: the result, using a simple spreadsheet, is shown in Figure 5.2(a) for eight rows at the top, middle, and bottom of the full 2^{16} rows. We now copy the *values* of the three-column table to three empty columns and then order the data according to the values of x. Figure 5.2(b) shows the result for 25 rows centered around $x = 1$. We have added two extra columns to the table in Figure 5.2(b), which show the result of subtracting the previous row exponents from the current row exponents. We have selected the rows above and below $x = 1$, because they all represent approximations to unity. Each of the 2-integers within a few rows of $x = 1$ represents a close

Output: Sequence of triples $\{a,b,x\}$.
procedure *1-digit-sequence(x)*

```
{
    a = b:= -128;
    if (a < 128) then do
    {
        if (b < 128) then do
        {
            x = 2^a 3^b; write (a,b,x);
            a:= a + 1;
            if (a = 128) then do;
            {
                    a:= -128; b:= b + 1;
            }
        }
    }
    else exit;
}
```

FIGURE 5.1
One-digit 2DLNS sequence generation algorithm.

approximation to unity, multiplication by which generates the next value in the sequence. Note that these have been naturally generated based on our constraints on the dynamic range of the exponents.

We observe the following:

1. The signs of the exponents in each row produce a fractional representation.
2. The values of x are inversely symmetrical above and below $x = 1$.
3. Apart from the above, there appears to be no easily definable ordering of the exponents.
4. The ratio of the values of x in adjacent rows is limited to a small number of different values, as evidenced by the exponents of the ratios in the fourth and fifth columns.
5. The values of the exponents of these ratios can exceed the 8-bit signed limit of the exponent dynamic range.

Observation 2 clearly sets this two-base system apart from the LNS. In the LNS there is a monotonic ordering of the exponents (LNS representation) with respect to the ordering of the values of the representation. The extra degree of freedom in this two-base system has, however, uncovered a potential problem related to the mapping of a number within a conventional representation (e.g., binary) to MDLNS and back again. This may seem to doom the representation, but we will address this in a specific application later in the chapter, with a full exposé of mapping techniques provided in the following chapters. We will find that this Achilles' heel is more than matched by the advantages of using the MDLNS in application-specific scenarios.

a	b	2^a3^b
−128	−128	2.493E-100
−128	−127	7.478E-100
−128	−126	2.243E-99
−128	−125	6.73E-99
−128	−124	2.0189E-98
−128	−123	6.0568E-98
−128	−122	1.8171E-97
−128	−121	5.4512E-97
....		
0	−3	0.03703704
0	−2	0.11111111
0	−1	0.33333333
0	0	1
0	1	3
0	2	9
0	3	27
0	4	81
....		
127	120	3.0575E+95
127	121	9.1724E+95
127	122	2.7517E+96
127	123	8.2551E+96
127	124	2.4765E+97
127	125	7.4296E+97
127	126	2.2289E+98
127	127	6.6867E+98

(a)

a	b	2^a3^b	$an - a_{n-1}$	$bn - b_{n-1}$
−73	46	0.93840007	−84	53
76	−48	0.94723872	149	−94
−8	5	0.94921875	−84	53
−92	58	0.95120292	−84	53
57	−36	0.96016215	149	−94
−27	17	0.96216919	−84	53
−111	70	0.96418043	−84	53
122	−77	0.97123172	233	−147
38	−24	0.9732619	−84	53
−46	29	0.97529632	−84	53
103	−65	0.98448249	149	−94
19	−12	0.98654037	−84	53
−65	41	0.98860255	−84	53
84	−53	0.99791405	149	−94
0	0	1	−84	53
−84	53	1.00209031	−84	53
65	−41	1.01152885	149	−94
−19	12	1.01364326	−84	53
−103	65	1.0157621	−84	53
46	−29	1.02532941	149	−94
−122	77	1.02962041	−84	53
111	−70	1.03715028	233	−147
27	−17	1.03931825	−84	53
−57	36	1.04149075	−84	53
92	−58	1.05130039	149	−94
8	−5	1.05349794	−84	53
−76	48	1.05570008	−84	53
73	−46	1.06564356	149	−94

(b)

FIGURE 5.2
Results from spreadsheet experiment.

Observation 3 is interesting, particularly when we look at the entire 2^{16} representations and find that only 14 different unity approximants are required for the 8-bit signed dynamic range of each of the exponents in the 2-integers. Observation 4 shows that exponents that exceed the 8-bit signed dynamic range are valid and useful providing that the result of adding these exponents to a valid 2-integer representation (i.e., within the specified dynamic range) also produces a valid representation. The values of these before and after representations will be close approximations to each other if a careful selection of the best unity approximant is made. Clearly this has happened automatically in generating the tables with the defined exponent dynamic range.

a	b	2^a3^b	Occurrence
−84	53	1.00209031	34,916
233	−147	1.00731325	2,507
149	−94	1.00941885	12,320
65	−41	1.01152885	8,904
−19	12	1.01364326	5,330
46	−29	1.02532941	456
27	−17	1.03931825	456
8	−5	1.05349794	456
−11	7	1.06787109	80
−3	2	1.125	80
5	−3	1.18518519	12
2	−1	1.33333333	12
−1	1	1.5	4
1	0	2	2

FIGURE 5.3
Ordered list of occurrences of the 14 unity approximants from 8-bit signed dynamic range exponents.

Figure 5.3 shows an ordered list of the 14 different unity approximants extracted from the 2^{16} 2-integers, ordered in ascending value of the unity approximant.

We note that the three unity approximants from Figure 5.2(b) are listed as the three closest approximations to unity in Figure 5.3. We also note that the inverses (opposite signs) of the approximants are also found in Figure 5.2(b) and are also available for use in calculations.

The usefulness of the existence of good approximations of unity, for general computations within dynamic constraints on the exponents, can be seen from the following example:

Example 5.1

Calculate x^2 by using 9-bit fixed-point arithmetic, where $x = (188, -120)$ in 2DLNS with odd base 3. The actual value of x is 0.218317.... Clearly, $x^2 = (376, -240)$, which would cause overflow in 9-bit arithmetic. However, if we multiply in advance by a (properly selected) good approximation of unity, then the result obtained will have much smaller binary and ternary exponents; consequently, there will not be any risk of overflow. In our case, if we multiply x by (−84, 53) we obtain (104, −67), and now the squaring can be achieved in 9-bit arithmetic without overflow.

More to the point, if at any stage of the computational process one obtains a pair of large exponents, they can be reduced to within the required exponent dynamic range by multiplying (adding the exponents of) the number obtained by a suitably good approximation of unity. *Note that this feature is not available for LNS, which has a single fixed base.*

5.4 Multiple-Digit MDLNS

When performing a computation using multidigit MDLNS, each digit can be treated as an independent MDLNS number and the operations handled separately. For example, if X and Y are two-digit MDLNS numbers such that $X = x_1 + x_2$ and $Y = y_1 + y_2$ then

$$X \cdot Y = (x_1 + x_2)(y_1 + y_2) = (x_1 \cdot y_1) + (x_1 \cdot y_2) + (x_2 \cdot y_1) + (x_2 \cdot y_2) \tag{5.7}$$

where x_i and y_i are single-digit MDLNS numbers. The independence of the arithmetic operations is very important, as it naturally allows for parallel architectures.

5.4.1 Error-Free Integer Representations

As discussed in Chapter 1, most often in DSP applications the input data have to be converted from analog to a fixed-point binary value with a uniform quantization error bound. Mapping to integers has a quantization error bounded by ±0.5 for all converted values. For a classical LNS representation (and also a one-digit MDLNS representation) we do not have this uniform quantization accuracy, so we have to choose a sufficient number of bits so that we will be able to maintain this conversion accuracy for the larger data values. In the multidigit MDLNS we can mitigate this quantization problem; in fact, we can find certain MDLNS representations that are completely error-free.

In the following, and as an example, consider the case of the two odd prime bases, (3, 5).

A representation of a real number into forms given in Definitions 5.3 to 5.5 is called *error-free* if the approximation error is zero. The next three theorems and one conjecture provide interesting results about the error-free two-dimensional logarithmic representation of numbers.

Theorem 5.1: Every real number, x, may have at most 14 different error-free two-digit two-dimensional logarithmic representations.

Proof: Let us assume that x is represented in the form of Definition 5.5:

$$x = \pm 2^a p^b \pm 2^c p^d \tag{5.8}$$

Clearly, x must be a rational number. Now we multiply the two sides of Equation (5.8) by $z = 2^{-\min(a,c,0)} p^{-\min(b,d,0)}$. The left and right sides of the new equation will be integers. Divide by the greatest common divisor of $(2^{a-\min(a,c,0)} p^{b-\min(b,d,0)}, 2^{c-\min(a,c,0)} p^{d-\min(b,d,0)})$. Let us denote the left side of the equation obtained as M. We may obtain only two types of equations:

$$M = \pm 1 \pm 2^e p^f \tag{5.9}$$

or

$$M = \pm 2^e p^f \tag{5.10}$$

Equation (5.9) may have at most one solution, due to the fundamental theorem of arithmetic. Equation (5.10) can be treated as follows. If the signs are different, we are in a position to apply the following result (recently proved by Bennett [10]): if a,b and c are integers, $a,b \geq 2$, then the equation $a^x - b^y = c$ may have at most two different solutions in integers (x, y). Therefore, Equation (5.10) may have, at most, four different solutions if the signs are different. If the signs are the same (say, positive, which corresponds to positive M), then we can do the following. Represent the exponents e and f with respect to modulo 3: $e = 3e_1 + e_2$ and $f = 3f_1 + f_2$, $e_2, f_2 \in \{0,1,2\}$. For the nine possible combinations of residues (e_2, f_2) we have nine Diophantine equations of the form

$$M = A2^{3e_1} + Bp^{3f_1} \tag{5.11}$$

where $A \in \{1,2,4\}$ and $B \in \{1, p, p^2\}$. We substitute $X = 2^{e_1}$ and $Y = p^{f_1}$ The final equation we have is

$$M = c_1 X^3 + c_2 Y^3 \tag{5.12}$$

where c_1 and c_2 are constants. Delone and Fadeev's theorem [11] about cubic Diophantine equations states that Equation (5.12) may have at most one solution in integers. Since we have nine Diophantine equations of this type, we may have at most nine different solutions of Equation (5.9).

Therefore, the total number of different error-free representations is bounded from above by 4 (the maximal number of solutions of Equation (5.10)) + 9 (the maximal number of solutions of Equation (5.11)) plus 1 (the maximal number of solutions of Equation (5.9)), that is, 14. ■

In some important special cases, this bound can be considerably improved.

Theorem 5.2: Every real number, x, may have at most seven different error-free two-digit two-dimensional logarithmic representations if the nonbinary base is 3.

Proof: Assume that x is represented in the form

$$x = \pm 2^a 3^b \pm 2^c 3^d \tag{5.13}$$

Proceeding in the same way as in the proof of Theorem 5.1, we obtain the following four equations:

$$M = \pm 1 \pm 2^e 3^f \tag{5.14}$$

$$M = 2^e - 3^f \tag{5.15}$$

$$M = -2^e + 3^f \tag{5.16}$$

$$M = 2^e + 3^f \tag{5.17}$$

Again, Equation (5.14) may have at most one solution in integers, and Equations (5.15) and (5.16) may have, at most, four (totally) different solutions in integers, according to Bennett's theorem [10]. Equation (5.17) can be treated by making use of the following theorem, due to Stroeker and Tijdeman [12]: all solutions of $2^x - 2^y = 3^z - 3^w$, $x > y > 0$, and $z > w > 0$ in integers x, y, z, w are given by (3, 1, 2, 1), (5, 3, 3, 1), and (8, 4, 5, 1). Therefore only three numbers (11, 35, and 259) may have two different representations as a sum of a power of 2 and a power of 3. That is, the total number of error-free representations is bounded from above by 7. ■

The upper bound, proved in Theorem 5.1, can certainly be improved. We have not found any real number with more than five error-free two-digit LNS representations, but here we report one case having exactly five error-free representations.

Let $x = 3.25$; then x can be represented with no error in a two-digit 2DLNS with odd base 3 in the following five expressions:

$$3.25 = \begin{cases} (1,-2,2,1,0,0) \\ (1,0,1,1,-2,0) \\ (1,2,0,-1,-2,1) \\ (1,1,1,-1,-2,2) \\ (1,6,0,-1,-2,5) \end{cases}$$

The point of Theorem 5.1 is to establish an effectively computable upper bound that can be a starting point for improvements. The example given shows that the lower bound for the maximal number of error-free representations is 5.

Theorem 5.3: The smallest positive integer with no error-free two-digit two-dimensional LNS representation, in the case of odd base 3, is 103.

Proof: The proof is based on the following result proved by Ellison [13]: let $x > 11$, $x \neq 13, 14, 16, 19, 27$; then for all $x, y \in N$ the following inequality holds:

$$|2^x - 3^y| > e^{x(\ln 2 - 0.1)}$$

Up to 102 we can give proper examples found by computational experiments. A simple check shows that 103 is not a sum of integers of the form 2^a3^b; 103 is not divisible by 2 or 3. Therefore it must be a difference of a power of 2 and a power of 3 (or vice versa). Applying Ellison's theorem we have

$$103 = |2^x - 3^y| > e^{x(\ln 2 - 0.1)} \tag{5.18}$$

Therefore x should be smaller than 8, and there are only 13 possible values for x, namely, $x \in \{1, 2, 3, 4, 5, 6, 7, 11, 13, 14, 16, 19, 27\}$. We now calculate and in none of the cases do we find that the corresponding y is an integer; therefore 103 cannot be represented in an error-free manner in a two-digit 2DLNS with bases 2 and 3. ■

Theorem 5.4: The smallest positive integer with no error-free two-digit two-dimensional logarithmic representation, in the case of nonbinary base 5, is 43.

Proof: The proof is based on a theorem, proved by Tijdeman [14], which has a much more general result than Ellison's result used in Theorem 5.3. Tijdeman states that if x and y are two consecutive s-integers, $s > 1$, then $|x - y| > y / (\log y)^c$, where c is an effectively computable constant. In the case of 2-integers, c is estimated to be less than 64 based on recent results in transcendental number theory [15]. We apply this theorem to numbers of the form 2^a5^b.

Again, up to 42 inclusively, we can find appropriate error-free representations. Forty-three is not a sum of two numbers of this form; therefore it could only be a difference, so we may apply Tijdeman's theorem and obtain an upper bound for the possible solutions. Namely, if $43 = |x - y|$, x and y being of the form 2^a5^b, then the theorem implies that $43(\log y)^{64} > y$; therefore $y < 2^{575}$. There are 296,371 numbers of the form 2^a5^b less than 2^{575}, that is, 296,371 potential values of x to be checked out. In none of these cases is the corresponding value for $x = y + 43$ a number of the form 2^a5^b. This shows that 43 is indeed the smallest positive integer without an error-free two-dimensional two-digit LNS representation for the case of nonbinary base 5. ■

We note that Tijdeman's theorem can be used for the proof of Theorem 5.2. We have, however, provided this proof based on a result concerning the set of bases 2 and 3, in order to point out the difficulties that may arise if one applies general theorems from transcendental number theory.

In Chapter 2 we provided the result that the equations $\pm 2^a 3^b \pm 2^c 3^d \pm 2^e 3^f = 4985$ do not have solutions in integers [16]. It is important to note that such results will be available (and different) for every particular set of bases that we choose. In this case (that is, a three-digit two-dimensional logarithmic representation with odd base 3) we see that a 12-bit error-free mapping is available, a useful dynamic range for many DSP applications. It should also be noted that the size of all of the exponents used (a, b, c, d, e, f) only requires 3-bit unsigned integers for their representation.

5.4.2 Non-Error-Free Integer Representations

Clearly, error-free representations are special cases of the MDLNS, but the extra degree of freedom provided by the use of multiple digits can mitigate the nonuniform quantization properties of the classical LNS.

To illustrate this, we present numerical results for mapping 10-bit signed binary input data to the two-digit 2DLNS where we treat the nonbinary base as a parameter. In order to demonstrate the ability to closely match input data with very small exponents, we have restricted the odd base exponent to 3 bits only. We are allowing the binary exponent to be unrestricted; however, due to the 10-bit input range, the system automatically limits itself to 6 bits. We will see in the next section that this has very little bearing on the overall complexity of the inner product implementation (i.e., the hardware complexity is mainly driven by the dynamic range of the nonbinary exponents). As stated above, we require quantization errors to be < 0.5 in order to match the quantization error of a binary representation. Table 5.1 shows the number of non-error-free representations along with the worst quantization error for nonbinary bases in the set {3, 5, 7, 11, 13, 17, 47}.

The goal of applying this approximation scheme is to reduce as much as possible the size of the nonbinary exponent(s). For example, with a nonbinary base of 47, $x = 0101001110_2 = 334_{10}$ is represented as

$$334_{10} = 0101001110_2 \rightarrow 2^9 47^{-1} + 2^{25} 47^{-3} = 334.082429$$

In this case we have used only three bits for the nonbinary exponents; that is, they are restricted to the set {−4, −3, −2, −1, 0, 1, 2, 3}. Although a base of

TABLE 5.1

Number of Non-Error-Free Representations and Worst Quantization Error for Different Bases

Base	3	5	7	11	13	17	47
Errors ≥ 0.5	10	58	18	20	4	6	2
Worst error	0.77778	1.232	0.7247	0.84091	0.9604	0.71443	0.5

Source: Muscedere, R., Dimitrov, V.S., Jullien, G.A. Multidimensional Logarithmic Number System. In *The VLSI Handbook*, W.-K. Chen (ed.), 2nd ed., pp. 84-1–84-35. Boca Raton, FL: CRC Press, 2006. With permission.

47 only has two non-error-free representations for a 10-bit signed range, it is possible to select a noninteger base that will provide completely error-free representations. We will explore this in more detail later in the book.

To compare these results with an implementation using a classical LNS representation, we need to determine the number of bits of the logarithm to produce an absolute error of < 0.5. A previous study [1] has found that we require $n + \log_2(n)$ bits for the logarithm in order to achieve this accuracy for an n-bit positive number [17]. We have, in fact, checked this for the case of $n = 9$ (used in our two-digit 2DLNS study), and 12 bits are required for the logarithm to satisfy the same accuracy. If we assume that the hardware complexity of the classical LNS representation is driven by the number of bits in the logarithm, then we can see a potential for an enormous reduction in the implementation complexity of the two-digit 2DLNS versus the classical LNS.

5.5 Half-Domain MDLNS Filter

The power of the MDLNS over the LNS is the ability to split the dynamic range of the single LNS exponent over two (or more) MDLNS exponents. The reduction of dynamic range allows the use of reasonably sized lookup tables, implemented in ROMs, to replace arithmetic logic. This hides the extra logic complexity of nonbinary number systems over their binary counterparts. For this first foray into a hardware implementation of the MDLNS, we make use of the mapping of multiplication to addition, but transform out of the MDLNS to binary when we need to perform the otherwise difficult operation of addition. A similar approach was first used by our group to implement index calculus hardware in a modulus replication residue number system (MRRNS) [18]. The basic idea is to use binary adders as multipliers in the logarithmic domain and then to use a ROM to map to the binary domain where binary adders are used for summing. This is a very specific approach that is limited to single cascaded multiplication paths in the flow graph. Fortunately, this is exactly the feature of the direct implementations of finite impulse response (FIR) filters. In fact, we can invoke the now standard inner product step processor that was originally pioneered for the systolic array implementation of convolution [19,20].

5.5.1 Inner Product Step Processor

Discrete convolution defines the mapping function of an input data stream to an output data stream for linear time-invariant systems. Aperiodic (noncircular) convolution, for a finite impulse response of length M samples, is given by Equation (5.19):

$$y_n = \sum_{m=0}^{M-1} h_m x_{n-m} \tag{5.19}$$

A companion function, discrete correlation, has the same structure as the convolution function, except that there is no time reversal of the input sequence; this is shown in Equation (5.20):

$$y_n = \sum_{m=0}^{M-1} h_m x_{n+m} \tag{5.20}$$

Both of these functions can be computed using a systolic inner product step processor (IPSP). Figure 5.4 shows a systolic correlator processor array for $M = 4$ along with the IPSP structure. The black squares represent latches, which are used to pipeline the data through the array. To demonstrate the action of the array, Table 5.2 shows the output of each step processor (SP) against 10 time steps. Note that the table entries are for the output of the step processors before the latches. The extra latch in the output data stream allows the input data to *slide across* the accumulating output data, which are required for the correlation function. The effect of the single latch in the input data stream can be observed at the beginning of the computation for each SP; the first nonzero output is delayed by a single time step for each step from the left to the right. The effect of the double delay in each SP on the accumulating output data stream can be seen in the partial computations

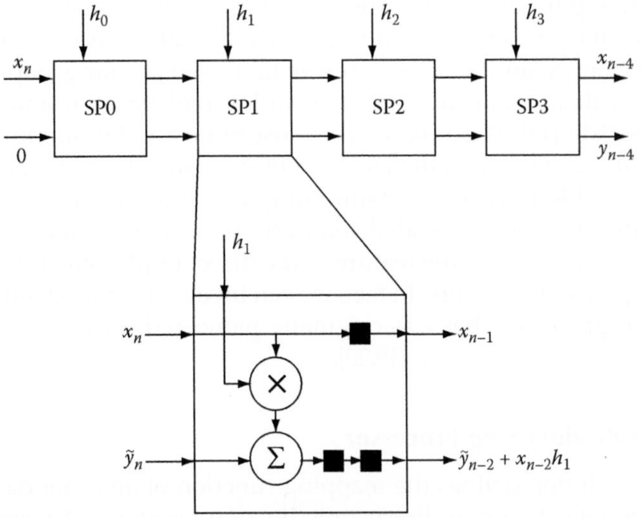

FIGURE 5.4
Systolic correlator for M = 4 using IPSP computational blocks.

TABLE 5.2

Detailed Computation for a Four-Step IPSP Correlator Array (outputs before the latches)

Time	SP0	SP1	SP2	SP3
0	x_0h_0			
1	x_1h_0	x_0h_1		
2	x_2h_0	$x_0h_0 + x_1h_1$	x_0h_2	
3	x_3h_0	$x_1h_0 + x_2h_1$	$x_0h_1 + x_1h_2$	x_0h_3
4	x_4h_0	$x_2h_0 + x_3h_1$	$x_0h_0 + x_1h_1 + x_2h_2$	$x_0h_2 + x_1h_3$
5	x_5h_0	$x_3h_0 + x_4h_1$	$x_1h_0 + x_2h_1 + x_3h_2$	$x_0h_1 + x_1h_2 + x_2h_3$
6	x_6h_0	$x_4h_0 + x_5h_1$	$x_2h_0 + x_3h_1 + x_4h_2$	$x_0h_0 + x_1h_1 + x_2h_2 + x_3h_3$
7	x_7h_0	$x_5h_0 + x_6h_1$	$x_3h_0 + x_4h_1 + x_5h_2$	$x_1h_0 + x_2h_1 + x_3h_2 + x_4h_3$
8	x_8h_0	$x_6h_0 + x_7h_1$	$x_4h_0 + x_5h_1 + x_6h_2$	$x_2h_0 + x_3h_1 + x_4h_2 + x_5h_3$
9	x_9h_0	$x_7h_0 + x_8h_1$	$x_5h_0 + x_6h_1 + x_7h_2$	$x_3h_0 + x_4h_1 + x_5h_2 + x_6h_3$

that are ringed in the table. For example, the $x_0h_0 + x_1h_1$ result at the output of SP1 (before the SP1 latches) in time step 2 does not appear at SP2, where it is summed with x_2h_2, until time step 4.

We can check the output from the table against Equation (5.20) for the ninth time step at the output, which represents a delayed version of the output from the first three steps of the input (note that there are six output latches between SP0 and the output of SP3, before the SP3 latches). The computation is shown in Equation (5.21).

$$y_3 = \sum_{m=0}^{3} h_m x_{m+3} = h_0x_3 + h_1x_4 + h_2x_5 + h_3x_6 \tag{5.21}$$

Convolution can be performed by reversing the $\{h\}$ sequence multipliers to the SPs or by feeding the accumulating output data in the reverse direction [20].

The following section details the design of a one-digit 2DLNS step processor in Figure 5.4.

5.5.2 Single-Digit 2DLNS Computational Unit

Our goal here is to use a special feature of the IPSP in that there is only a single multiplier in any signal path from the input to the output. This means that

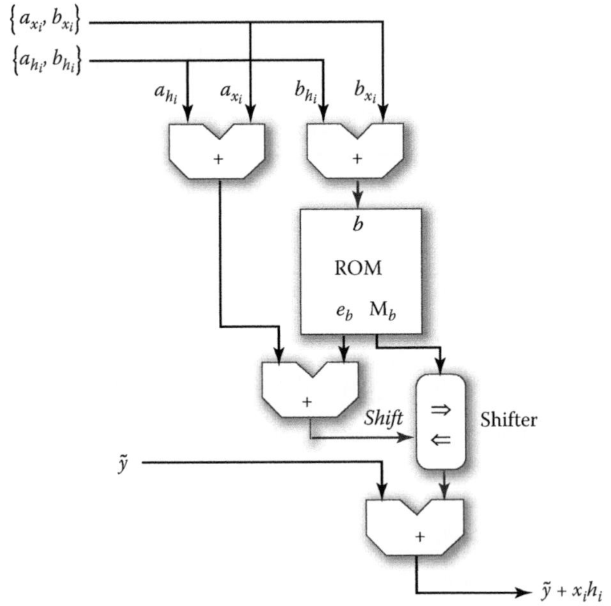

FIGURE 5.5
One-digit 2DLNS step processor. (From Dimitrov, V.S., Jullien, G.A., Miller, W.C. Theory and Applications of the Double-Base Number System. *IEEE Transactions on Computers* 48(10) (1999): 1098–1106. With permission.)

we can use the MDLNS to perform the multiplications, immediately followed by a conversion to binary. We will use the fact that the first base is binary to simplify this procedure. The entire two-base SP is shown in Figure 5.5. We will use the notation from Definition 5.5, $x_i = 2^{a_{xi}} 3^{b_{xi}}$ and $h_i = 2^{a_{h_i}} 3^{b_{h_i}}$, giving the 2DLNS representations $x_i \Rightarrow \{a_{x_i}, b_{x_i}\}$ and $h_i \Rightarrow \{a_{h_i}, b_{h_i}\}$. For simplicity we have removed the sign bit from the representations and the hardware descriptions. The difficult problem of converting to an MDLNS from binary is covered in later chapters. For low-dynamic-range DSP applications (such as low-precision video processing), a simple lookup table stored in a reasonably sized ROM will represent a suitable solution. In this case the interest is in computing over a larger dynamic range than that of the input data, as discussed in Chapter 1.

The conversion to binary, following the addition of the exponents, uses a two-step procedure: (1) conversion to a binary floating point form, and (2) conversion to a binary fixed-point form. The steps arise naturally from the fact that the first base in the 2DLNS representation is 2. If we convert the ternary exponent, *b*, of a 2DLNS representation, {*a*, *b*}, to a floating point form, $3^b \approx 2^e \cdot M$, then we can simply add the exponent, *e*, to the exponent of the binary base to yield the floating point form, $2^{e+a} \cdot M$, where M· is the mantissa. The second step in the conversion only requires a shifter in which the mantissa is shifted by the amount determined by $e + a$. This is exactly the same procedure that is used in floating point addition to align the exponents of the addends before performing a standard binary addition. The output from the shifter is finally summed with the partially accumulated output, $\tilde{y}$, in a standard binary adder. The sign bits from x_i and h_i can be handled separately and used to control the output adder of each SP (not shown).

The conversion of 3^b takes place in a ROM using precomputed values for the exponent and mantissa. The accuracy required for the mantissa can be determined from a noise analysis, as discussed in Chapter 1. The size of the ROM grows exponentially with the number of bits used for *b*, and only linearly with the accuracy requirements. Searching for small dynamic range ternary exponents is thus a useful exercise.

As an example of generating the ROM contents, consider the case of a 4-bit range for *b*, the ternary exponent. We will use a two's complement coding for the exponent, which will provide eight positive values (including zero) and eight negative values. We choose 10 as the number of binary digits of precision in the fractional part of the mantissa.

Table 5.3 shows the entire procedure. The input to the ROM is the ternary exponent, *b* (column 1 from the left). The value represented by this exponent is 3^b (column 2). We now find the exponent, e^b, shown in column 3. The exponent is chosen so that $1 \le 3^b \cdot 2^{-e^b} < 2$, which guarantees a normalized mantissa. Column 4 contains $3^b \cdot 2^{(10-e_b)}$, rounded to an integer, which converts the normalized mantissa range to $2^{10} \le 3^b \cdot 2^{(10-e_b)} < 2^{11}$ and provides the 10 bits of precision as required. Column 4 contains the 11 binary digits of the value in column 3 with the binary point to the right of the most significant bit (a 1). This position of the binary point is equivalent to dividing by 2^{10}, bringing the range of the mantissa back to $1 \le M_b < 2$. Finally, column 5 shows the reconstruction of the stored values using the 10-bit fractional part of the mantissa. We note a few small errors in some of the values. It is this error that we take into account when choosing the precision of the mantissa. The relevant columns for the programming of the ROM use bold font: these are columns 1, 3, and 5.

A rather unusual situation occurs with the generation of the exponent, $shift = a_{b_i} + a_{x_i} + e_b$, for the shifter as shown in Figure 5.5. In particular, the exponents, $\{a_{h_i}, a_{x_i}\}$, can have large changes in value (positive and negative) for even very small differences in the value of the 2DLNS representation (see the first column of Table 5.4). However, in typical correlation or convolution applications, the *shift* input to the shifter will be much smaller than the maximum right or left shift dictated by the individual exponents of x_i and h_i. In fact, we can independently perform an analysis of the dynamic range experienced by the array for typical signals in order to place bounds on shift, without even considering how the computations are to be performed. This is a very unusual situation, since an independent (and a priori) knowledge of the dynamic range of the *shift* input means that we can

TABLE 5.3

Detailed Table Calculations for the ROM Contents of Figure 5.2

b	3^b	e_b	$M_b \cdot 210$	M_b	$\tilde{3}^b$
0	1	0	1,024	1.0000000000	1
1	3	1	1,536	1.1000000000	3
2	9	3	1,152	1.0010000000	9
3	27	4	1,728	1.1011000000	27
4	81	6	1,296	1.0100010000	81
5	243	7	1,944	1.1110011000	243
6	729	9	1,458	1.0110110010	729
7	2,187	11	1,094	1.0001000110	2,188
–8	0.00015	–13	1,279	1.0011111111	0.00015
–7	0.00046	–12	1,918	1.1101111110	0.00046
–6	0.00137	–10	1,438	1.0110011110	0.00137
–5	0.00412	–8	1,079	1.0000110111	0.00412
–4	0.01235	–7	1,618	1.1001010010	0.01234
–3	0.03704	–5	1,214	1.0010111110	0.03705
–2	0.11111	–4	1,820	1.1100011100	0.11108
–1	0.33333	–2	1,365	1.0101010101	0.33325

invoke the observation in Chapter 1; i.e., we only need to implement two's complement adders of sufficient dynamic range to hold the maximum right and left shift inputs to the shifter. This means that the word length of the components of shift only has to have as many bits as the maximum shift range; this will bring an attendant reduction in storage, wiring, and addition hardware.

As an example of this reduced word length effect, we have taken a snapshot of a SP within a simulation of a typical FIR filter (see Figure 5.6), where we have determined that the shift range of values are *shift* $\in \{-16, +16\}$. This corresponds to a Mod(2^5) reduction that can be made to all calculations in the shift chain. In effect, for a two's complement binary representation, all inputs to calculations only need to use (and store) their five least significant bits, and two's complement adders can be built with only a 5-bit input and output word length. We have not included the accumulator to simplify the diagram.

In this snapshot, the coefficient has the value $h_i = 5 \rightarrow \{-69, 45\}$, and the data sample has the value $x_i = 29 \rightarrow \{-263, 169\}$. Note, in Figure 5.7, that since the binary exponents only operate on the shifter, they have been reduced: Mod(2^5). The ROM contains the ternary exponent lookup

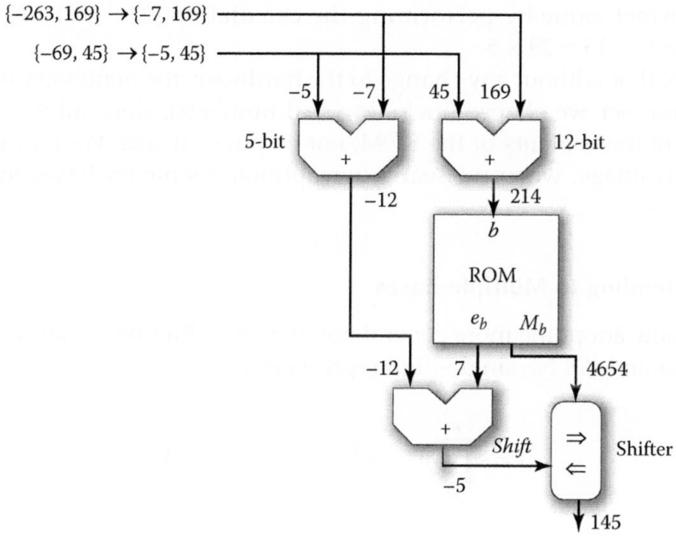

FIGURE 5.6
Snapshot of a SP computation.

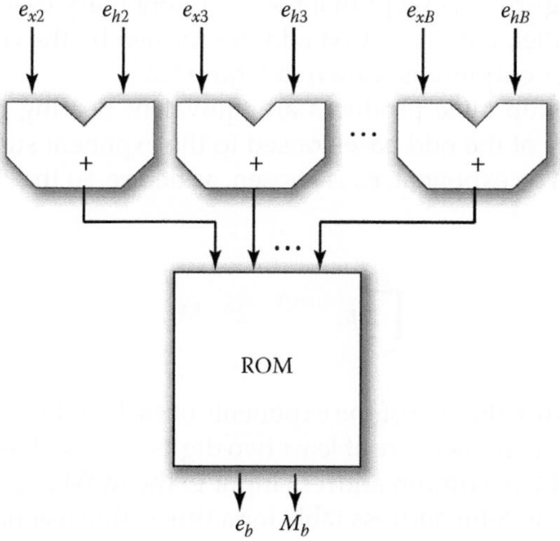

FIGURE 5.7
ROM lookup for a B-base MDLNS.

table from which we obtain the result $3^{214} \approx 4654.2^{327}$. This is stored as $\{e_b, M_b\} \rightarrow \{7{,}4654\}$, again with the reduction 327 Mod(2^5) = 7 of the binary exponent, since this only contributes to the shift value. The output of the shift computation is −5–7 + 7 = −5, and the right shift of 4,654 by five places yields the shifted truncated result of 145. We can verify this as the correct value by performing the calculation $(2^{-263} \cdot 3^{169}) \times (2^{-69} \cdot 3^{45}) = 2^{-332} \cdot 3^{214} \approx 145 = 29 \times 5$.

We note that without any change to the hardware, the nonbinary base can be any number we choose (including real numbers), since all we need to change are the contents of the ROM, not its physical size. We will use this to our advantage, when we search for optimal nonbinary bases for given applications.

5.5.3 Extending to Multiple Bases

We will now adopt the more general notation of Definition 5.2 with simplifications for an MDLNS single-digit representation:

$$x = s\prod_{j=1}^{B} p_j^{e_j} \Rightarrow \{s, e_{x1}, e_{x2}, \cdots, e_{xB}\} \tag{5.22}$$

where x is represented by the B-tuple of exponents, together with the sign bit, as shown in Equation (5.22). As with the 2DLNS, $p_1 = 2$; however, we have already shown that the second base need not be an integer for a 2DLNS, and so in general, $\{p_2, \ldots, p_B\}$ may not be integers. The SP structure is almost identical to that of Figure 5.5, except that the $B - 1$ nonbinary base exponents are summed in parallel, with the ROM address formed by the concatenation of the parallel adder outputs, as shown in Figure 5.7.

This ROM lookup table produces an equivalent floating point value for the product of all of the odd bases raised to the exponent sum, as shown in Equation (5.23). The exponent, e_b, is chosen, as before, so that the mantissa is normalized.

$$\prod_{j=2}^{B} p_j^{(e_{xj}+e_{hj})} \approx 2^{e_b} \cdot \mathrm{M}_b \tag{5.23}$$

We note that since the size of the exponents of each odd base in an MDLNS representation (where there are at least two digits and two bases) can be very small (< 4 bits), the maximum address input to the ROM is given by $4 \cdot (B - 1)$ bits. This is only an 8-bit address table for a three-dimensional MDLNS.

For MDLNS representations with a dimension of > 2, we can also consider the use of unity approximants to reduce the output of each odd base adder to the number of bits of the input exponents (or even less if we are willing to accept the increased mapping error). This reduction process stores a small number of unity approximants (see Figure 5.3 for a 16-bit 2DLNS system) that can be added in parallel to the output of the odd base adders. The reduced input to the ROM is selected from these parallel results. The ROM input address size is now reduced by $(B - 1)$ bits.

Table 5.4

Sorting the SP ROM Contents of Table 5.3 by Mantissa Value

M_b·210	M_b	e_b	b
1,024	1.0000000000	0	0
1,079	1.0000110111	–8	–5
1,094	1.0001000110	11	7
1,152	1.0010000000	3	2
1,214	1.0010111110	–5	–3
1,279	1.0011111111	–13	–8
1,296	1.0100010000	6	4
1,365	1.0101010101	–2	–1
1,438	1.0110011110	–10	–6
1,458	1.0110110010	9	6
1,536	1.1000000000	1	1
1,618	1.1001010010	–7	–4
1,728	1.1011000000	4	3
1,820	1.1100011100	–4	–2
1,918	1.1101111110	–12	–7
1,944	1.1110011000	7	5
2,048	1.0000000000	–1	0

5.5.4 Extending to Multiple Digits

We have provided an overview of a two-digit MDLNS multiplication in Equation (5.7), where the four partial products of the multiplication can be computed in parallel with a final addition of the partial products to produce the output result. In general the two-digit computational array is a simple parallel extension of the one-digit array. Each of the units computes the binary output for one of the digit combinations. As an example, consider, in detail, the two-digit MDLNS multiplication, $z = x \cdot y$, where

$$y = \sum_{i=1}^{2} s_i^{[y]} \prod_{j=2}^{B} p_j^{e_j^{[y](i)}} \; ; \; x = \sum_{i=1}^{2} s_i^{[x]} \prod_{j=2}^{B} p_j^{e_j^{[x](i)}}$$

We can perform this multiplication with four parallel one-digit multipliers, where the (u, v) multiplier computes

$$z_{u,v} = s_u^{[y]} \cdot s_v^{[x]} \prod_{j=2}^{B} p_j^{\left(e_j^{[y](u)} + e_j^{[x](v)}\right)} \tag{5.24}$$

Clearly there are n^2 such multipliers in an n-digit MDLNS multiplier.

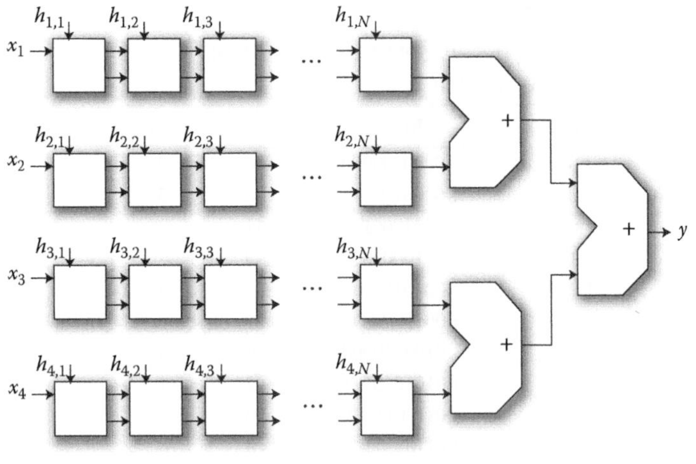

FIGURE 5.8
Overall structure of a two-digit MDLNS correlator.

This concept can be extended to IPSP architectures, where parallel partial data streams are processed with their outputs being summed at the end of the systolic arrays using an adder tree. The overall structure using a two-digit MDLNS, for both the $\{x\}$ and $\{h\}$ sequences, is shown in Figure 5.8. Based on the clock speed limit for pipelining the SP blocks, we may be able to compute the 3 adder tree within one clock cycle, which will reduce the pipeline latency by one clock pulse.

The main advantage of the use of more than one digit for the input data and the filter coefficients is that one can obtain extremely accurate representations with very small nonbinary exponents. The price that has to be paid, however, is that the number of computational channels required is increased to at least four. One possible way to reduce the number of computational channels in a two-digit MDLNS is to use a greedy algorithm to find the digits. The greedy technique finds a representation with the first digit that is as close as possible to the number being represented. The second digit is used to minimize the error. With this approach, the first digit will represent a much larger magnitude than the second. If we do this for both the $\{x\}$ and $\{h\}$ sequences, then the fourth computational channel only involves multiplication and additions with the much smaller magnitude second digits. We may be able to remove this fourth channel if the error introduced is acceptable for the application.

5.5.5 General Binary-to-MDLNS Conversion

Whatever number representation is used in a computational procedure, the input data will usually be in a binary form. In the previous section, we assumed that the initial conversion from binary to MDLNS took place in a ROM, where the conversion exponents were stored at the address given by the input binary data. For small-input word lengths, a ROM lookup table (LUT) is probably the best approach in terms of hardware cost on silicon. The word length limit to this lookup table approach depends on the technology, and we do have the advantage in forward conversion that only a single ROM is required to drive a potentially large computational array. However, the LUT size is driven not only by the input word length but also by the number of digits and bases in the MDLNS representation. For example, an error-free (where the absolute representation error is less than 0.5) 12-bit unsigned range with three digits and two bases would require a direct mapping LUT of size 4096×33 or 136 Kbit, a reasonably sized Very large-scale integration (VLSI) component. However, if an error-free 23-bit unsigned range were needed, the mapping LUT would be 8388608×48 or 403 Mbit, which is not reasonable at all for current VLSI fabrication technologies.

In Chapter 6 we explore techniques and such trade-offs, as illustrated above, related to the problem of general MDLNS conversion. These techniques are used in the processor architectures introduced in Chapters 7–9.

5.6 Summary

This chapter has provided an initial exploration of the multidimensional logarithmic number system (MDLNS), which is based on the index calculus introduced in Chapter 2. The generalization of this representation to the MDLNS is formed by allowing more than two orthogonal bases and more than one digit, and we have provided definitions to formalize the representation.

A most important part of this chapter has been the introduction to implementing arithmetic. The MDLNS has the same easy multiplication/division and difficult addition/subtraction characteristics as the logarithmic number system (LNS), all based on the manipulation of exponents of the radix rather than the positional number representation itself. Our implementation solutions follow those of the LNS, albeit with multiple exponents, and we highlight the obvious advantages in reduction of lookup table size with smaller dynamic range exponents.

We have explored with examples, and some mathematical rigor, two interesting characteristics of the MDLNS, namely, approximations to unity and error-free multidigit representations. The usefulness of unity approximants (they do not exist in the classical LNS) allows the equivalent of scaling in the MDLNS, and multiple digits provide low-error (including error-free) mappings of input integers.

This chapter has also discussed the implementation of a systolic 2DLNS FIR filter, in which the advantages of easy multiplication within the representation are exploited. Summation, a difficult operation in the 2DLNS, is computed by local conversion to binary at each of the processors in the array. Some interesting observations arise from the construction of the single-step processor, including the ability to reduce the dynamic range of the finite ring arithmetic based on a priori knowledge of the filter characteristics, rather than using a much larger dynamic range that will not overflow during the calculations.

A final note on the necessity to find general conversion techniques into the MDLNS representation along with efficient addition and subtraction algorithms within the MDLNS leads us to the next four chapters on MDLNS hardware techniques.

References

1. Arnold, M.G., Bailey, T.A., Cowles, J.R., Cupal, J.J. Redundant Logarithmic Arithmetic. *IEEE Transactions on Computers* 39(8) (1990): 1077–1086.
2. Kingsbury, N.G., Rayner, P.J.W. Digital Filtering Using Logarithmic Arithmetic. *Electronics Letters* 7 (1971): 56–58.
3. Swartzlander, E.E., Alexopoulos, A.G. The Sign/Logarithm Number System. *IEEE Transactions on Computers* 37 (1975): 1238–1242.
4. Taylor, F.J., Gill, R., Joseph, J., Radke, J. A 20 Bit Logarithmic Number System Processor. *IEEE Transactions on Computers* 37(8) (1988): 190–200.
5. Muller, J.M., Scherbyna, A., Tisserand, A. Semi-Logarithmic Number Systems. *IEEE Transactions on Computers* 47(2) (1998).
6. Dimitrov, V.S., Jullien, G.A., Miller, W.C. Theory and Applications of the Double-Base Number System. *IEEE Transactions on Computers* 48(10) (1999): 1098–1106.
7. Lewis, D.M. 114 MFLOPS Logarithmic Number System Arithmetic Unit for DSP Applications. *IEEE Journal of Solid-State Circuits* 30 (1995): 1547–1553.
8. Coleman, J.N., Chester, E.I., Softley, C.I., Kadlec, J. Arithmetic on the European Logarithmic Microprocessor. *IEEE Transactions on Computers* 49(7) (2000): 702–715.
9. Muscedere, R. Difficult Operations in the Multi-Dimensional Logarithmic Number System. PhD thesis, University of Windsor, Ontario, Canada, 2003.
10. Bennett, M.A. On Some Exponential Equation of S.S. Pillai. *Canadian Journal of Mathematics* 53 (2001): 897–922.
11. Delone, B.N., Fadeev, D.K. The Theory of Irrationalities of the Third Degree. *Translation of Mathematical Monographs* 10 (1964).
12. Stroeker, R.J., Tijdeman, R. Diophantine Equations, Computational Methods in Number Theory (Part II). In *MC Tracts*, R.H.W. Lenstra Jr. and R. Tijdeman (eds.), vols. 154–155, pp. 321–369. Amsterdam: Mathematical Centre, 1983.
13. Ellison, W.J. *On a Theorem of Sivasankaranayana—Seminars on Number Theory*. Technical Report 12. CNRS, Talene, 1971.
14. Tijdeman, R. On the Maximal Distance between Integers Composed of Small Primes. *Compositio Mathematica* 28(2) (1974): 159–162.

15. Waldshmidt, M. Linear Independence Measures for Logarithms of Algebraic Numbers. In *International Mathematical Year*, pp. 1–92. Cetraro, 2000.
16. Dimitrov, V., Howe, E. Lower Bounds on the Length of Double-Base Representations. *Proceedings of the American Mathematical Society* 10 (2011): 3405–3411.
17. Muscedere, R., Dimitrov, V.S., Jullien, G.A., Miller, W.C. Efficient Techniques for Binary to Multi-Digit Multi-Dimensional Logarithmic Number System Conversion Using Range Addressable Look-Up Tables. *IEEE Transactions on Computers* 54(3) (2005): 257–271.
18. Jullien, G.A., Luo, W., Wigley, N.M. Hight Throughput VLSI DSP Using Replicated Finite Rings. *Journal of VLSI Signal Processing* 14 (1996): 207–220.
19. McCanny, J.V., McWhirter, J.G. Optimized Bit Level Systolic Array for Convolution. *IEE Proceedings* 131(Pt. G, no. 6) (1984): 632–637.
20. Kung, S.Y. *VLSI Array Processors*. Englewood Cliffs, NJ: Prentice-Hall, 1988.
21. Muscedere, R., Dimitrov, V.S., Jullien, G.A. Multidimensional Logarithmic Number System. In *The VLSI Handbook*, W.-K. Chen (ed.), 2nd ed., pp. 84-1–84-35. Boca Raton, FL: CRC Press, 2006.

6

Binary-to-Multidigit Multidimensional Logarithmic Number System Conversion

6.1 Introduction

The previous chapter introduced the multidimensional logarithmic number system (MDLNS) and discussed the relatively simple operations of multiplication and division as well as the difficult operations of addition, subtraction, and conversion to standard representations. In this chapter we will look in detail at novel, efficient techniques for implementing conversion between the MDLNS and standard binary representations. The techniques we discuss will also be applicable to the implementation of the other difficult MDLNS operations, such as addition and subtraction, which will be discussed in Chapter 7. The main contents of this chapter are based on a 2005 and a 2007 IEEE publication by the authors [1,2].

A LNS representation, a, of the number, x, is given by the relationship in Equation (6.1), where s is the sign of the number and the base is r (usually 2).

$$x = s \cdot r^{a} \tag{6.1}$$

As discussed in Chapter 5, the MDLNS provides more degrees of freedom than the LNS by virtue of the orthogonal bases and the ability to gain from the use of multiple digits. However, these extra degrees of freedom introduce new complexities in the binary conversion process. The binary-to-LNS conversion process is simplified due to the monotonic relationship between x and a. Unfortunately, this solution is not applicable to the MDLNS since there is no monotonic relationship between x and the multiple digits/bases.

The technique initially proposed for binary-to-MDLNS conversion used simple lookup tables (LUTs) [3]. Although a LUT offers a simple and fast binary-to-MDLNS conversion scheme, the size of the LUT can become very large and impractical since their size is exponentially dependent on the input binary dynamic range. In fact, the original application area for the index calculus DBNS was in video stream processing, where the input has a relatively small dynamic range (on the order of 8 bits). The LUT sizes further depend on the number of digits and bases in the MDLNS representation. For example, an error-free (where the absolute representation error is less than 0.5) 12-bit unsigned range with three digits and two bases [4] would require a direct mapping LUT of the size 4096 × 33 or 136 kbit, a reasonably sized component. However, if an error-free 23-bit unsigned range were needed, the mapping LUT would be 8388608 × 48 or 403 Mbit, which is not reasonable at all for hardware implementation.

This chapter will explore several hardware realizable techniques for converting a binary representation into a two-dimensional logarithmic number system (2DLNS) representation. The techniques are based on the reversal of the MDLNS-to-binary converter discussed in Chapter 5, with the aid of a new memory device.

To simplify the presentation of the binary-to-MDLNS process we will restrict ourselves to a subset of the MDLNS with only two bases (an n-digit 2DLNS representation), and we will assume that the exponent of the second (or nonbinary) base has a predefined finite precision equivalent to limiting the number of bits of precision in a classical LNS. The simplified representation of an input, x, as an n-digit 2DLNS is shown in Equation (6.2):

$$x = \sum_{i=1}^{n} s_i \cdot 2^{a_i} \cdot D^{b_i} \tag{6.2}$$

The second base, D, is a suitably chosen number (relatively prime to 2), $s_i = \{-1, 0, +1\}$, and the exponents are integers. R is the constrained precision of the second base exponent (i.e., bit width: $b_i = \{-2^{R-1}, \ldots, 2^{R-1}-1\}$), and it directly affects the complexity of the MDLNS system. We shall also define B as the constrained precision of the binary exponent (i.e., bit width: $a_i = \{-2^{B-1}, \ldots, 2^{B-1}-1\}$). Unlike R, B only weakly affects the complexity of the system. We define these values since our MDLNS system is to be realized in hardware.

6.2 Single-Digit 2DLNS Conversion

We start our explanation by looking at the single-digit 2DLNS case. Setting $n = 1$ in Equation (6.2), we obtain the single-digit 2DLNS representation:

$$x = s \cdot 2^a \cdot D^b \tag{6.3}$$

6.2.1 Single-Digit 2DLNS-to-Binary Conversion

We first review the process for converting from a single-digit 2DLNS to binary [3]. This method uses b as an index address to a LUT to find a floating point representation for D^b, as shown in Equation (6.4).

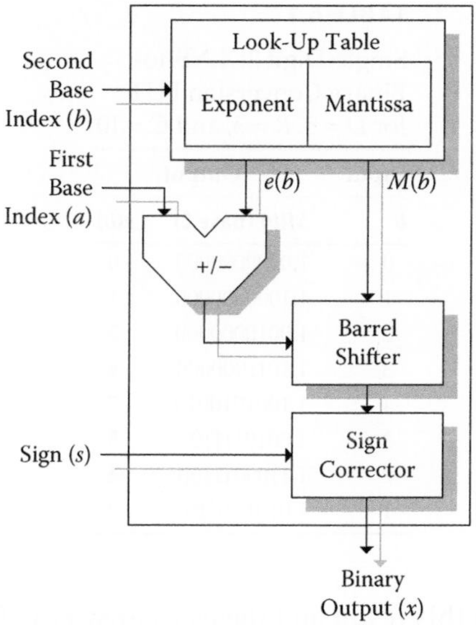

FIGURE 6.1
Single-digit 2DLNS-to-binary converter structure. (From R. Muscedere, V. Dimitrov, G.A. Jullien, W.C. Miller, Efficient Techniques for Binary-to-Multidigit Multidimensional Logarithmic Number System Conversion Using Range-Addressable Look-Up Tables, *IEEE Transactions on Computers*, 54, 257–271, 2005. Copyright © 2005 IEEE. With permission.)

$$D^b = M(b) \cdot 2^{e(b)} \tag{6.4}$$

Here $M(b)$ is the mantissa ($1 \le M(b) < 2$) and $e(b)$ is the exponent (integer). The final floating point representation of x is shown in Equation (6.5), and the conversion architecture in Figure 6.1.

$$x = s \cdot M(b) \cdot 2^{(a+e(b))} \tag{6.5}$$

Table 6.1 shows the single-digit 2DLNS-to-binary conversion table for $D = 3$ and $R = 3$. The number of table rows is 2^R (8 for this example). The precision of the mantissa is $C = 10$, where C is the number of bits of the fractional part.

To demonstrate this conversion process, we provide the following example:

Example 6.1

Given s = +1, $a = 8$, and $b = -3$, find x using Equation (6.5) and Table 6.1.

Solution

- Look up $b = -3$ in Table 6.1.
- Table returns $M(-3) = 1.0010111101_2$ and $e(-3) = -5$.
- Therefore $x = +1 \times 1.0010111101_2 \times 2^{(8-5)} = 9.4765625$ or 256/27.

TABLE 6.1
Single-Digit 2DLNS-to-Binary Conversion LUT for $D = 3$, $R = 3$, and $C = 10$

Input	Output	
b	$M(b)$ (base 2)	$e(b)$
0	1.0000000000	0
1	1.1000000000	1
2	1.0010000000	3
3	1.1011000000	4
–4	1.1001010010	–7
–3	1.0010111101	–5
–2	1.1100011100	–4
–1	1.0101010101	–2

The error between this result and the exact answer of 9.481481… is due to the precision of the mantissa ($C = 10$).

6.2.2 Binary-to-Single-Digit 2DLNS Conversion

Since the table method for converting single-digit 2DLNS to binary is quite fast and can be implemented efficiently in hardware, it is only logical to try to reverse this process, i.e., to convert a binary two's complement representation to a single-digit 2DLNS. For this conversion, the sign of the binary input, x, must first be determined and the magnitude, $|x|$, generated. If x is in a two's complement format, the high-order (or sign) bit is used to determine the sign (unless $|x|$), and $x = 0$ is generated from the two's complement of x. If x is in a floating point (FP) format (i.e., IEEE 754), the sign is determined from the FP sign bit (unless $x = 0$, a special FP case). In the special event where $s = 0$, the 2DLNS exponents will also be set to 0 in order to minimize the chance of arithmetic overflow when performing simple 2DLNS operations (multiplication, division, etc.).

The input to the single-digit 2DLNS to binary LUT is the second base exponent, b, and the outputs are the mantissa and the exponent. To reverse the process, the input to the LUT is the mantissa, $M(b)$. Since the mantissa is not influenced by the exponent, this exponent can remain an output. Table 6.2 shows preliminary binary-to-single-digit-2DLNS LUT contents for $D = 3$ and $R = 3$. If we were to implement this LUT in hardware, the complexity of the nonshaded area would be $O(2^C)$; however, since the table contains undefined entries for all possible input values, aside from the 2^R values that are exactly representable (shown as ↓ and referred to as ranges), the complexity is really only $O(2^R)$. The shaded table entries, downward, are formed by doubling the mantissas and decrementing the exponents by 1. Shaded table

TABLE 6.2

Preliminary Binary-to-Single-Digit 2DLNS Conversion LUT for $D = 3$, $R = 3$

Input	Output	
$M(b)$	$e(b)$	b
...	...	...
↓	?	?
0.8888888888	−3	−2
↓	?	?
1.0000000000	0	0
↓	?	?
1.1250000000	3	2
↓	?	?
1.1851851851	−5	−3
↓	?	?
1.3333333332	−2	−1
↓	?	?
1.5000000000	1	1
↓	?	?
1.5802469145	−7	−4
↓	?	?
1.6875000000	4	3
↓	?	?
1.7777777777	−4	−2
↓	?	?
2.0000000000	−1	0
↓	?	?
...	...	...

entries, upward, have mantissas halving and exponents incrementing by 1. Our complexity expression excludes any calculations based on the output bit widths since they are considered a constant.

In order to reduce the LUT complexity to $O(2^R)$ (i.e., 2^R ranges) we need to remove the undefined entries and restrict the input range to that of the normalized mantissa (the nonshaded area in Table 6.2). The latter is achieved by the external LUT address generation hardware; the removal of the undefined entries can be achieved by rounding any input in an undefined range to the nearest representable value, i.e., using a midpoint function between the defined addresses. We therefore have the complete LUT for $D = 3$, $R = 3$, and $C = 10$, as shown in Table 6.3. Note that an extra range is required due to the possibility of rounding up numbers near 2.0. Thus the number of ranges is $2^R + 1$ but the complexity is maintained at $O(2^R)$.

TABLE 6.3

Complete Binary-to-Single-Digit 2DLNS Conversion LUT for $D = 3$, $R = 3$, and $C = 10$

Input	Output	
$M(b)$ (base 2)	**$e(b)$**	**b**
1.0000000000 ↓ 1.0000111111	0	0
1.0001000000 ↓ 1.0010011101	3	2
1.0010011110 ↓ 1.0100001000	–5	–3
1.0100001001 ↓ 1.0110101001	–2	–1
1.0110101010 ↓ 1.1000101000	1	1
1.1000101001 ↓ 1.1010001000	–7	–4
1.1010001001 ↓ 1.1011101101	4	3
1.1011101110 ↓ 1.1110001101	–4	–2
1.1110001110 ↓ 1.1111111111	–1	0

The mapping of $|x|$, in the two's complement case, to its equivalent mantissa is easily achieved in hardware with a conditional feedback bit shifter and counter (multicycle process, lower power), or priority encoder (single-cycle process, higher power). In either case, the number of shifts performed, *shifts*, is used to generate the binary exponent, as shown in Equation (6.6):

$$a = shifts - e(\text{mantissa}) \tag{6.6}$$

In the case of floating point, $|x|$ is already represented in a compatible mantissa form. The number of shifts can be determined from the exponent portion of the FP notation. However, this method is only valid for normalized

FP notations. Denormalized FP notations (i.e., values less than 2^{-126} for 32-bit FP and 2^{-1022} for 64-bit FP, including zero) will require additional hardware as in the two's complement case. We will limit our discussion to normalized FP notations only.

The following example demonstrates the use of Table 6.3 in performing a two's complement binary-to-single-digit 2DLNS conversion.

Example 6.2

Given $x = 1392$, find s, a, and b.

Solution

- Since $x > 0$, $s = +1$.
- Normalize $|x| = 1392$ to $|x| = 1.359375 \times 2^{10}$ or $|x| = 1.010111_2 \times 2^{10}$ (10 shifts).
- Find the mantissa 1.010111_2 in the LUT (Table 6.3).
- The LUT returns $b = -1$ with an exponent of -2.
- $a = 10 - (-2) = 12$.
- Therefore, $+1 \times 2^{12} \times 3^{-1} = +1365$ is the nearest approximation with $R = 3$

Unfortunately, in order to build a conventional LUT (similar to Table 6.3), all possible values of the mantissa must be accommodated in the address decoder, which increases the complexity to $O(2^{\varsigma})$.

6.3 Range-Addressable Lookup Table (RALUT)

In this section we discuss the design of a novel conversion LUT with table size $O(2^R)$. In this approach we change the address decode system from exact matching to range matching. We will refer to this as a range-addressable LUT (RALUT).

6.3.1 RALUT Architecture

A standard LUT architecture is shown in Figure 6.2(a), where an address decoder is used to match the address to a unique stored value. The RALUT architecture of Figure 6.2(b) shows the new address decoder system that matches a stored value to a range of addresses. The decoder compares the input address, I, to a range of two neighboring monotone addresses (e.g., Addr(1) and Addr(2)). Only one of these comparisons will match the input and activate a word-enable line that drives the data patterns, *Data*, to the output, O, of the RALUT.

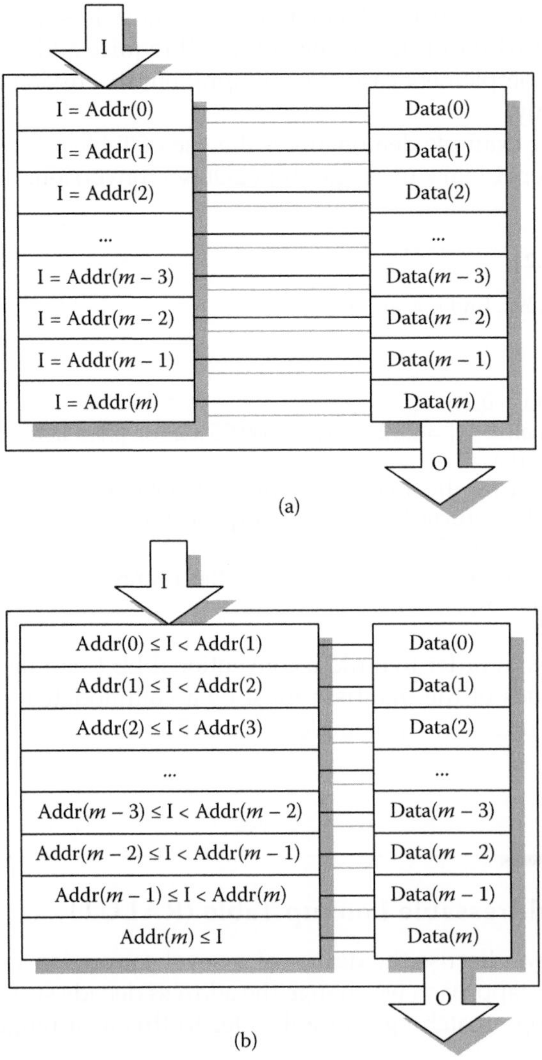

FIGURE 6.2
Standard and range-addressable LUT architectures. (From R. Muscedere, V. Dimitrov, G.A. Jullien, W.C. Miller, Efficient Techniques for Binary-to-Multidigit Multidimensional Logarithmic Number System Conversion Using Range-Addressable Look-Up Tables, *IEEE Transactions on Computers*, 54, 257–271, 2005. Copyright © 2005 IEEE. With permission.)

We can remove half of the comparators in the range decoder by noting that

$$(I < Addr(n)) = \overline{(Addr(n) \le I)} \tag{6.7}$$

$$(I \ge Addr(n)) \oplus (I \ge Addr(n+1)) = Addr(n) \le I < Addr(n+1) \tag{6.8}$$

Since the addresses are monotonic, if $(I \ge Addr(n + 1))$ is true, then $(I \ge Addr(n))$ must also be true. Therefore we can reduce the XOR operator and rewrite Equation (6.8) as

$$(I \ge Addr(n)) \cdot \overline{(I \ge Addr(n+1))} = Addr(n) \le I < Addr(n+1) \tag{6.9}$$

The optimized uniform architecture is shown in Figure 6.3. Further gate level optimizations can be made (i.e., last AND gate, removal of first comparator, etc.); however, in order to keep the explanation simple, they will not be shown here.

The contents of the RALUT for $D = 3$, $R = 3$, and $C = 10$ are shown in Table 6.4. The input column contains the values for which $(I \ge Addr(n))$ match (beginning range of each row in Table 6.3), and the output columns contain the associated outputs. The RALUT architecture is optimal since it only requires $2^R + 1$ rows.

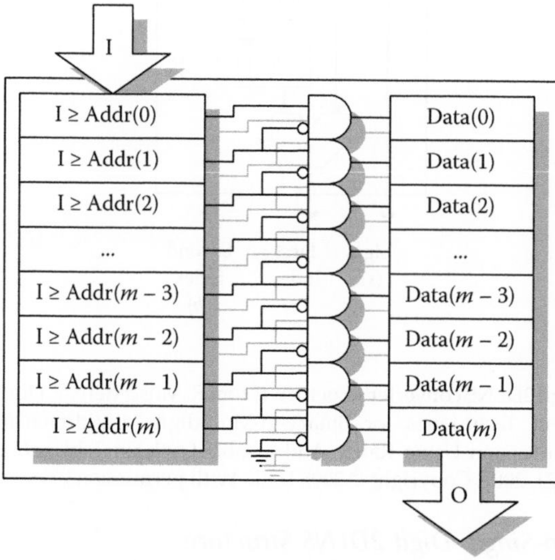

FIGURE 6.3
Logically optimized RALUT structure. (From R. Muscedere, V. Dimitrov, G.A. Jullien, W.C. Miller, Efficient Techniques for Binary-to-Multidigit Multidimensional Logarithmic Number System Conversion Using Range-Addressable Look-Up Tables, *IEEE Transactions on Computers*, 54, 257–271, 2005. Copyright © 2005 IEEE. With permission.)

TABLE 6.4

Complete Binary-to-Single-Digit 2DLNS Conversion RALUT for $D = 3$, $R = 3$, and $C = 10$

Input	**Output**	
***M*(*b*) (base 2)**	***e*(*b*)**	***b***
1.0000000000	0	0
1.0001000000	3	2
1.0010011110	–5	–3
1.0100001001	–2	–1
1.0110101010	1	1
1.1000101001	–7	–4
1.1010001001	4	3
1.1011101110	–4	–2
1.1110001110	–1	0

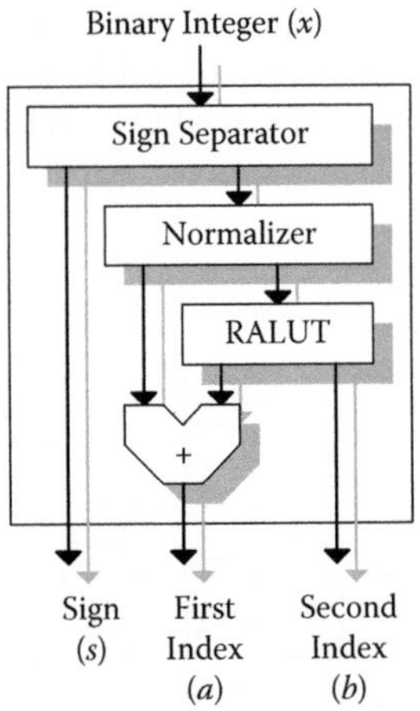

FIGURE 6.4
Binary-to-single-digit 2DLNS converter structure. (From R. Muscedere, V. Dimitrov, G.A. Jullien, W.C. Miller, Efficient Techniques for Binary-to-Multidigit Multidimensional Logarithmic Number System Conversion Using Range-Addressable Look-Up Tables, *IEEE Transactions on Computers*, 54, 257–271, 2005. Copyright © 2005 IEEE. With permission.)

6.3.1.1 Binary-to-Single-Digit 2DLNS Structure

The complete binary-to-single-digit 2DLNS structure using the RALUT is shown in Figure 6.4. This structure can be implemented in a single (low-latency) or multiprocess (pipelined, higher-latency, lower-power) circuit depending on the system constraints.

6.4 Two-Digit 2DLNS-to-Binary Conversion

From Equation (6.3), we obtain the two-digit 2DLNS representation shown in Equation (6.10).

$$x = s_1 \cdot 2^{a_1} \cdot D^{b_1} + s_1 \cdot 2^{a_2} \cdot D^{b_2} \tag{6.10}$$

Conversion from a two-digit 2DLNS-to-binary representation is a fairly simple process. Both 2DLNS digits are converted separately using the single-digit method and their results accumulated to produce the final binary representation. An example of this procedure is shown below.

Example 6.3

Given $s_1 = -1$, $a_1 = 5$, $b_1 = 1$, $s_2 = +1$, $a_1 = 2$, and $b_2 = 0$, find x.

Solution

- For x_1, look up $b_1 = 1$ from Table 6.1.
- Table 6.1 returns $M(1) = 1.1_2$ and $e(1) = 1$
- $x_1 = -1 \times 1.1_2 \times 2^{5+1} \times = -1100000_2$.
- For x_2, look up $b_2 = 0$ from Table 6.1.
- Table 6.1 returns $M(0) = 1.0_2$ and $e(0) = 0$.
- $x_2 = 1 \times 1.0_2 \times 2^{2+0} = 100_2$.
- Add x_1 and x_2.
- The final value is $x = x_1 + x_2 = -1011100_2$ or -92.

6.4.1 Two-Digit 2DLNS-to-Binary Conversion Architecture

The architecture for this conversion can be either serial or parallel. For the serial implementation we perform a two-cycle single-digit 2DLNS-to-binary conversion and use a single LUT as shown in Figure 6.5(a). The parallel structure uses two single-digit 2DLNS-to-binary conversion LUTs but only requires a single-cycle operation, as shown in Figure 6.5(b).

6.5 Binary-to-Two-Digit 2DLNS Conversion

In [3], simple LUT tables were used to convert a binary two's complement input into a MDLNS representation. The tables were generated by computing all possible MDLNS representations given the full range of B and R and selecting the nearest integer values. This method is obviously not feasible for

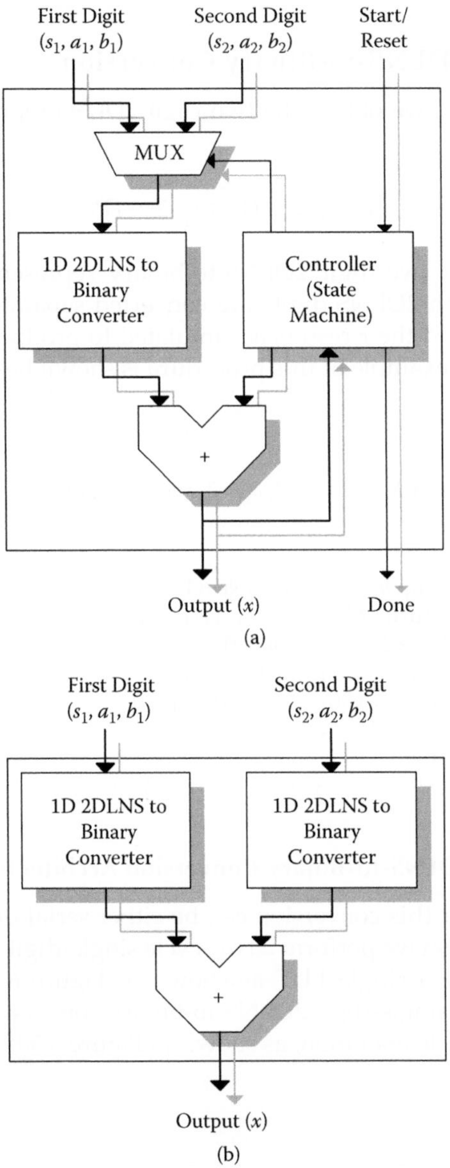

FIGURE 6.5
Serial and parallel two-digit 2DLNS-to-binary converter architectures. (From R. Muscedere, V. Dimitrov, G.A. Jullien, W.C. Miller, Efficient Techniques for Binary-to-Multidigit Multidimensional Logarithmic Number System Conversion Using Range-Addressable Look-Up Tables, *IEEE Transactions on Computers*, 54, 257–271, 2005. Copyright © 2005 IEEE. With permission.)

a real-time hardware solution. To address this issue, we have developed four methods for multidigit MDLNS conversion:

1. The *quick method* chooses the nearest first digit to the target and generates the second digit to reduce the error, a simple greedy algorithm.
2. The *high/low method* chooses the two nearest approximations to the target as the first digits, generates two associated second digits for the error, and selects the combination with the smaller error.
3. The *brute-force method* operates by selecting the combination with the smallest error, but it uses all possible mantissas of D^b as the first digits instead of just one (quick) or two (high/low).
4. The *extended-brute-force method* improved upon the brute-force method by using first-digit approximations above 2.0 and below 1.0 (shifted left or right L bits).

Each method ranges from simple implementations and fairly accurate approximations to difficult implementations and very accurate approximations.

We will compare the accuracy of each conversion method through their maximum relative representation error (MRRE) and average relative representation error (ARRE) [5] at the end of this section. The relative representation error (RRE) is defined in Equation (6.11):

$$\frac{F(x)-x}{x} \tag{6.11}$$

where x is the target value for conversion and $F(x)$ is the nearest approximation. The MRRE is the largest possible error given all x's and ARRE is the average error for all x's.

6.5.1 Binary-to-Two-Digit 2DLNS Conversion (Quick Method)

A two-digit approximation provides many different possibilities to represent a real number in 2DLNS. The only technique we have found to obtain the best two-digit approximation is an exhaustive search, but a very good approximation can be found using the following simple greedy algorithm, greedy in the sense that it takes as much of the value as possible at each step.

Algorithm 6.1: Quick Method for MDLNS

Input: Real number, x.
Output: $\{s_1, a_1, b_1\}$, $\{s_2, a_2, b_2\}$ such that $x \approx s_1 \cdot 2^{a_1} \cdot D^{b_1} + s_2 \cdot 2^{a_2} \cdot D^{b_2}$.

Step 1: Find $\{s_1, a_1, b_1\}$ based on the binary-to-single-digit 2DLNS conversion of x.

Step 2: Generate $\tilde{x} = s_1 \cdot M(b_1) \cdot 2^{a_1+e(b_1)}$ using an additional output of the RALUT from step 1.
Step 3: Determine the error $x - \tilde{x}$ using a matched mantissa output (matched by exponent) from the RALUT (see Table 6.5).
Step 4: Find $\{s_2, a_2, b_2\}$ based on a binary-to-single-digit 2DLNS conversion of $x - \tilde{x}$.

In order to determine the error for the second-digit approximation, we see that the RALUT output for the first-digit approximation must also include the matched mantissa (see Table 6.5 and Example 6.4). The size of the RALUT is therefore increased due to this extra output, but it is a linear, not exponential growth. This method of binary-to-two-digit 2DLNS conversion will also be referred to as the quick method to determine a two-digit 2DLNS approximation; it uses minimal resources and is fast.

The following example illustrates the greedy algorithm technique.

Example 6.4

Given $x = 3840$, find s_1, a_1, b_1, s_2, a_2, and b_2.

Solution

- Since $x = 3840 > 0$, $s_1 = +1$.
- $|x| = 3840$.
- Normalize $|x|$.
- $|x| = 1.875 \times 2^{11}$ or $|x| = 1.111_2 \times 2^{11}$.
- Find 1.111_2 in the RALUT (Table 6.5).
- $b_1 = -2$ with $e = -4$ and $M = 01.11000111_2$.
- $a_1 = 11 - (-4) = 15$.
- Matching mantissa error: $(1.111_2 - 01.11000111_2) \times 2^{11} = 0.00011001_2 \times 2^{11}$

TABLE 6.5

Complete Binary-to-Single-Digit 2DLNS Conversion RALUT with Mantissa Output for $D = 3$, $R = 3$, and $C = 10$

Input	Output		
M(b) (base 2)	**e(b)**	**b**	**M(b) (base 2)**
1.0000000000	0	0	01.0000000000
1.0001000000	3	2	01.0010000000
1.0010011110	–5	–3	01.0010111101
1.0100001001	–2	–1	01.0101010101
1.0110101010	1	1	01.1000000000
1.1000101001	–7	–4	01.1001010010
1.1010001001	4	3	01.1011000000
1.1011101110	–4	–2	01.1100011100
1.1110001110	–1	0	10.0000000000

Second Digit

- $x = 0.00011001_2 \times 2^{11}$, since $x > 0$ $s_2 = +1$, and $|x| = 0.00011001_2 \times 2^{11}$.
- Normalized, $|x|$ is $|x| = 1.1001_2 \times 2^7$ (7 shifts).
- From the RALUT (Table 6.5) $b_2 = -4$ with $e = -7$ and $M = 1.100101_2$.
- $a_2 = 7 - (-7) = 14$.

Final Approximation

- $+1 \times 2^{15} \times 3^{-2} + 1 \times 2^{14} \times 3^{-4} = 3843$.

This two-digit 2DLNS RRE is $\frac{3840 - 3843}{3840} = 0.00078125$.

6.5.1.1 *Quick Binary-to-Two-Digit 2DLNS Conversion Architecture*

For the second-digit approximation, the mantissa output of the RALUT is not required. In a serial hardware implementation of this converter, only a single RALUT (with mantissa output) is needed. The first-digit approximation uses the mantissa output while the second-digit approximation ignores it, as in Figure 6.6(a). A pipelined hardware implementation of this converter can take advantage of this by using two separate RALUTs (one with the mantissa output and one without) to minimize the area, as in Figure 6.6(b).

6.5.2 Binary-to-Two-Digit 2DLNS Conversion (High/Low Method)

Although the quick binary-to-two-digit 2DLNS conversion method operates efficiently, it does not always provide the most accurate result. Depending on the choice of R, there can be many possible representations for a number. The quick method allows the choice of the first digit to be above or below the target value, based on the nearest approximation, but independently of the selection of the second digit. An alternative method, the high/low method, will generate two representations, one with the first digit below the target and the other with the first digit above the target. In either case, the choice of the two first digits is the two nearest approximations to the target, a modified greedy algorithm. Once the second digit is generated (from the error calculated from the first digit and the mantissa in the RALUT), then the more accurate of the two two-digit approximations can be chosen. Note that one of these final representations will be the same as that produced by the quick method. To determine the best representation, the second RALUT will also require the mantissa output so that the error from the second digit can be calculated. We formalize this technique in Algorithm 6.2.

Algorithm 6.2: High/Low Method for MDLNS

Input: Real number, x.

Output: $\{s_1, a_1, b_1\}$, $\{s_2, a_2, b_2\}$ such that $x \approx s_2 \cdot 2^{a_1} \cdot D^{b_1} + s_2 \cdot 2^{a_2} \cdot D^{b_2}$.

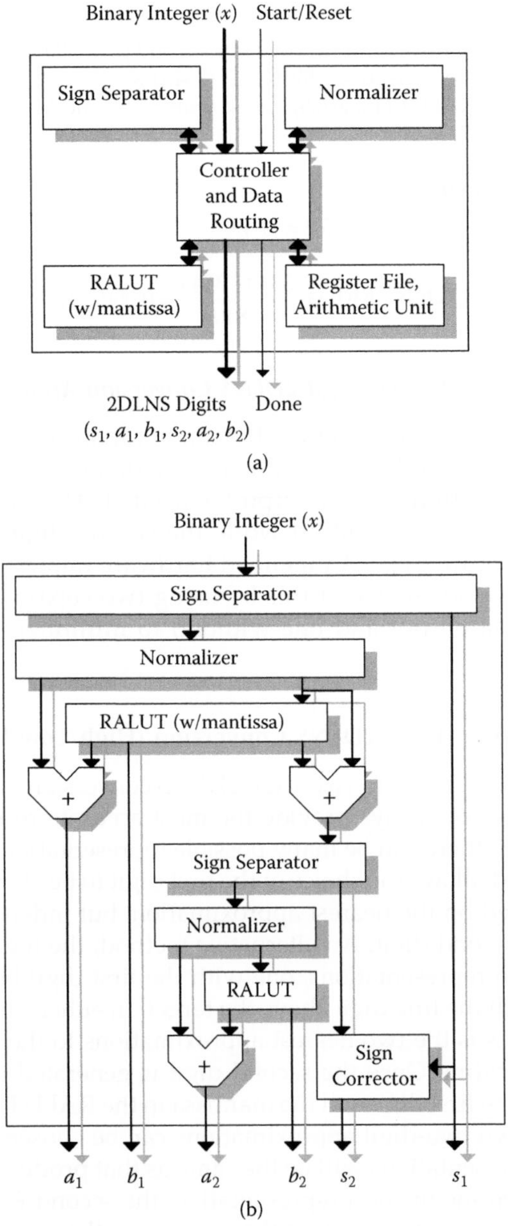

FIGURE 6.6
Quick binary-to-two-digit 2DLNS converter architectures. (From R. Muscedere, V. Dimitrov, G.A. Jullien, W.C. Miller, Efficient Techniques for Binary-to-Multidigit Multidimensional Logarithmic Number System Conversion Using Range-Addressable Look-Up Tables, *IEEE Transactions on Computers*, 54, 257–271, 2005. Copyright © 2005 IEEE. With permission.)

Step 1: Find $\{s_1^-, a_1^-, b_1^-\}$ and $\{s_1^+, a_1^+, b_1^+\}$ based on the binary-to-single-digit 2DLNS conversion of $x^- \le x$ and $x^+ > x$, respectively.
Step 2: Generate $\tilde{x}^- = s_1^- \cdot M(b_1^-) \cdot 2^{a_1^- + e(b_1^-)}$ and $\tilde{x}^+ = s_1^+ \cdot M(b_1^+) \cdot 2^{a_1^+ + e(b_1^+)}$ using additional outputs of the RALUT from step 1.
Step 3: Determine the errors $x - \tilde{x}^-$ and $x - \tilde{x}^+$ using the matched mantissa output (matched by exponent) from the RALUT (see Table 6.6).
Step 4: Find $\{s_2, a_2, b_2\}$ based on a binary-to-single-digit 2DLNS conversion of $x - \tilde{x}^-$ or $x - \tilde{x}^+$, which minimizes the error.

6.5.2.1 *Modifying the RALUT for the High/Low Approximation*

In order to implement the high/low method, the RALUT has to store both the high and low approximations for the first digit. Since this method will test both the high and low approximations for the first digit, the RALUT addresses are modified to include the range from the low to the high value of each row. The RALUT contents for $D = 3$, $R = 3$, and $C = 10$ are shown in Table 6.6. Note that the high contents of each row are equal to the low contents of the next row. The number of table rows is 2^R, one row less than the previous tables, since cyclical connectivity is maintained by the high element of the last row instead of requiring a new row.

The following example illustrates the high/low binary-to-two-digit 2DLNS conversion method.

TABLE 6.6

Binary-to-Single-Digit 2DLNS Conversion RALUT for High/Low Method for $D = 3$, $R = 3$, and $C = 10$

Input	Output					
	Low			High		
M(*b*) (base 2)	*e*(*b*)	*b*	*M*(*b*) (base 2)	*e*(*b*)	*b*	*M*(*b*) (base 2)
1.0000000000	0	0	01.0000000000	3	2	01.0010000000
1.0010000000	3	2	01.0010000000	−5	−3	01.0010111101
1.0010111101	−5	−3	01.0010111101	−2	−1	01.0101010101
1.0101010101	−2	−1	01.0101010101	1	1	01.1000000000
1.1000000000	1	1	01.1000000000	−7	−4	01.1001010010
1.1001010010	−7	−4	01.1001010010	4	3	01.1011000000
1.1011000000	4	3	01.1011000000	−4	−2	01.1100011100
1.1100011100	−4	−2	01.1100011100	−1	0	10.0000000000

Example 6.5

Given $x = 3840$, find s_1, a_1, b_1, s_2, a_2, and b_2 for the most accurate representation using the high/low method.

Solution

- Since $x = 3840 > 0$; $s_1 = +1$ and $|x| = 3840$.
- Normalize $|x|$: $|x| = 1.875 \times 2^{11}$ or $|x| = 1.111_2 \times 2^{11}$.
- Find 1.111_2 in the RALUT (Table 6.6).

Low Approximation, First Digit

- $b_1 = -2$ with $e = -4$ and $M = 01.11000111_2$
- $a_1 = 11-(-4) = 15$
- Matching mantissa error: $(1.111_2 - 01.11000111_2) \times 2^{11} = 0.00011001_2 \times 2^{11}$

Low Approximation, Second Digit

- $x = 0.00011001_2 \times 2^{11}$, therefore $s_2 = +1$ and $|x| = 0.00011001_2 \times 2^{11}$
- Normalized, $|x|$ is $|x| = 1.1001_2 \times 2^7$ (7 shifts)
- From the RALUT (Table 6.5) $b_2 = -4$ with $e = -7$ and $M = 1.00101_2$
- $a_2 = 7-(-7) = 14$
- Find the difference (or approximation error) between the two mantissas: $(1.1001_2 - 1.100101_2) \times 2^7 = 0.000001_2 \times 2^7$ or $1.0_2 \times 2^1 + 1 \times 2^{15} \times 3^{-2} + 1 \times 2^{14} \times 3^{-4} = 3843$

  with a computed error of 2
- Same solution as quick method

High Approximation, First Digit

- $b_1 = 0$ with $e = -1$ and $M = 10.0000000000_2$
- $a_1 = 11-(-1) = 12$
- Matching mantissa error: $(1.111_2 - 10.0_2) \times 2^{11} = -0.001_2 \times 2^{11}$

High Approximation, Second Digit

- $x = -0.001_2 \times 2^{11}$, therefore $s_2 = -1$ and $|x| = 0.001_2 \times 2^{11}$
- Normalized, $|x|$ is $|x| = 1.0_2 \times 2^8$ (8 shifts)
- From the RALUT (Table 6.5) $b_2 = 0$ with $e = 0$ and $M = 01.0_2$
- $a_2 = 8-(0) = 8$
- Find the difference (or approximation error) between the two mantissas: $(1.0_2 - 1.0_2) \times 2^8 = 0 + 1 \times 2^{12} \times 3^0 - 1 \times 2^8 \times 3^0 = 3840$

  with a computed error of 0
- First digit not greedy

Final Approximation

- $+1 \times 2^{12} \times 3^0 - 1 \times 2^8 \times 3^0 = 3840$.

6.5.2.2 *High/Low Binary-to-Two-Digit 2DLNS Architecture*

For a pipelined hardware implementation, a single RALUT with dual outputs (as in Table 6.6) and two RALUTs with mantissa output are needed (see Figure 6.7). For a serial hardware implementation, it is possible to use only one RALUT (dual outputs). The input mantissa can be compared with the two output mantissas to determine the nearest approximation for the second digit. The implementation is similar to that of the quick method (see Figure 6.6(b)) with some additional resources (storage and arithmetic).

6.5.3 Binary-to-Two-Digit 2DLNS Conversion (Brute-Force Method)

The high/low conversion method offers more approximations than the quick method at the expense of more hardware. For the parallel implementation,

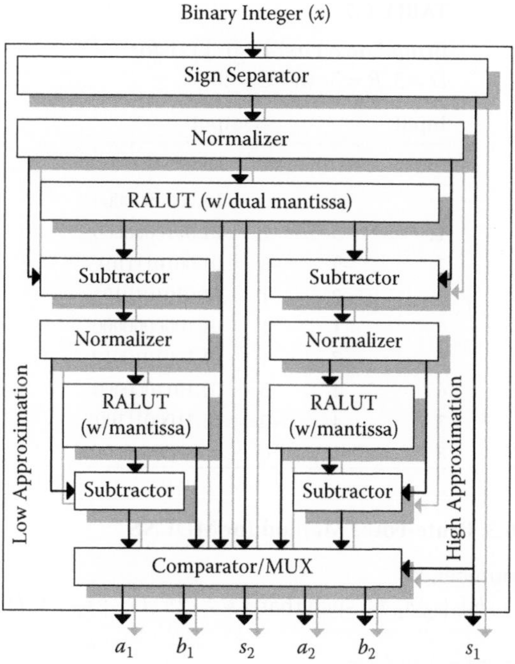

FIGURE 6.7
High/low binary-to-two-digit 2DLNS parallel converter structure. (From R. Muscedere, V. Dimitrov, G.A. Jullien, W.C. Miller, Efficient Techniques for Binary-to-Multidigit Multidimensional Logarithmic Number System Conversion Using Range-Addressable Look-Up Tables, *IEEE Transactions on Computers*, 54, 257–271, 2005. Copyright © 2005 IEEE. With permission.)

a second RALUT is required. However, for the serial implementation, the resources remain almost the same. Depending on the particular application for which 2DLNS is used, it may be necessary to have even more accurate two-digit 2DLNS representations. For such cases it is possible to perform a brute-force two-digit 2DLNS conversion.

The quick method chooses the nearest first digit to the target and generates the second digit to reduce the error, a simple greedy algorithm. The high/low method chooses the two nearest approximations to the target as the first digits, generates two associated second digits for the error, and selects the combination with the smaller error. The brute-force method operates similarly by selecting the combination with the smallest error, but it uses all possible mantissas of D^b as the first digits instead of just one (quick) or two (high/low).

A LUT (see Table 6.7) containing all the mantissa and exponent information is required to be accessed to generate each of the combinations $(2^R + 1)$. The best combination is determined iteratively. The brute-force method is formally described in Algorithm 6.3, and illustrated in Example 3.6.

TABLE 6.7

Brute-Force First-Digit LUT for $D = 3$, $R = 3$, and $C = 10$

Input	Output		
Index	$e(D^b)$	b	$M(D^b)$ **(base 2)**
0	0	0	01.0000000000
1	3	2	01.0010000000
2	–5	–3	01.0010111101
3	–2	–1	01.0101010101
4	1	1	01.1000000000
5	–7	–4	01.1001010010
6	4	3	01.1011000000
7	–4	–2	01.1100011100
8	–1	0	10.0000000000

Algorithm 6.3: Brute-Force Method for MDLNS

Input: Real number, x.
Output: $\{s_1, a_1, b_1\}$, $\{s_2, a_2, b_2\}$ such that $x \approx s_1 \cdot 2^{a_1} \cdot D^{b_1} + s_2 \cdot 2^{a_2} \cdot D^{b_2}$.

Step 1: Set $i = 0$ and $\varepsilon = \infty$.
Step 2: Find $\{s_1^i, a_1^i, b_1^i\}$ from row i of a LUT of all the mantissa values.
Step 3: Generate $\tilde{x}^i = s_1^i \cdot M(b_1^i) \cdot 2^{a_1^i + e(b_1^i)}$ and determine $\{s_2, a_2, b_2\}$ based on minimizing the error $\varepsilon_i = |x - (\tilde{x}_1^i + \tilde{x}_2^i)|$.
Step 4: Set $i \to i + 1$ and $\varepsilon = \min(\varepsilon, \varepsilon_i)$ and iterate from step 2 until all LUT rows have been searched.
Step 5: Final output is $\{s_1^j, a_1^j, b_1^j\}$, $\{s_2^j, a_2^j, b_2^j\}$, where j corresponds to the row that the smallest error, ε, is obtained.

Example 3.6

Given $x = 1967$, find s_1, a_1, b_1, s_2, a_2, and b_2 using the brute-force method.

Solution

- Quick method results in $+1 \cdot 2^{11} \cdot 3^0 - 1 \cdot 2^8 \cdot 3^1 = 1962.\overline{6}$.
- High/low method results in $+1 \cdot 2^{14} \cdot 3^{-2} + 1 \cdot 2^4 \cdot 3^2 = 1964.\overline{4}$.
- Using Table 6.7 for all the entries:

  0: $+1 \cdot 2^{10} \cdot 3^0 + 1 \cdot 2^{13} \cdot 3^{-2} = 1934.\overline{2}$

  1: $+1 \cdot 2^7 \cdot 3^2 + 1 \cdot 2^{16} \cdot 3^{-4} = 1961.0865$

  2: $+1 \cdot 2^{15} \cdot 3^{-3} + 1 \cdot 2^8 \cdot 3^1 = 1981.\overline{629}$

  3: $+1 \cdot 2^{12} \cdot 3^{-1} + 1 \cdot 2^{14} \cdot 3^{-3} = 1972.\overline{148}$

  4: $+1 \cdot 2^9 \cdot 3^1 + 1 \cdot 2^4 \cdot 3^3 = 1968.0$

5: $+1 \cdot 2^{17} \cdot 3^{-4} + 1 \cdot 2^{10} \cdot 3^{-1} = 1959.506$

6: $+1 \cdot 2^{6} \cdot 3^{3} + 1 \cdot 2^{11} \cdot 3^{-2} = 1955.\overline{5}$

7: $+1 \cdot 2^{14} \cdot 3^{-2} + 1 \cdot 2^{4} \cdot 3^{2} = 1964.\overline{4}$

8: $+1 \cdot 2^{11} \cdot 3^{0} - 1 \cdot 2^{8} \cdot 3^{1} = 1962.\overline{6}$

- Entry 4 contains the nearest approximation to 1967.

Final Approximation

- $+1 \cdot 2^{9} \cdot 3^{1} + 1 \cdot 2^{4} \cdot 3^{3} = 1968.0$

6.5.3.1 Brute-Force Conversion Architecture

The *brute-force method* of two-digit 2DLNS conversion is primarily intended for a serial implementation since the process requires using the RALUT $2^R + 1$ times. A state machine (similar to the one used in the quick method of Figure 6.6(a)) can easily coordinate the movement of data to perform this task. It is possible to integrate both the LUT (for the first digit) and the RALUT (for the second digit) into a single structure in order to reduce area as both tables have the same output (see Figure 6.8). For a parallel implementation, a considerable amount of hardware is required, including $2^R + 1$ RALUTs, to maintain the data flow from one pipeline stage to another. Such an implementation is not recommended.

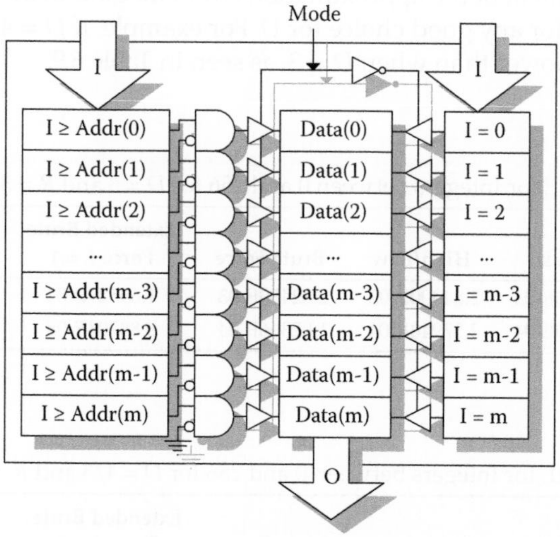

FIGURE 6.8
Combination LUT and RALUT structure. (From R. Muscedere, V. Dimitrov, G.A. Jullien, W.C. Miller, Efficient Techniques for Binary-to-Multidigit Multidimensional Logarithmic Number System Conversion Using Range-Addressable Look-Up Tables, *IEEE Transactions on Computers*, 54, 257–271, 2005. Copyright © 2005 IEEE. With permission.)

6.5.4 Binary-to-Two-Digit 2DLNS Conversion (Extended-Brute-Force Method)

The accuracy of the two-digit 2DLNS brute-force conversion method can be improved by using first-digit approximations above 2.0 and below 1.0. This can be implemented by shifting the mantissa after it has been retrieved from the LUT and prior to being subtracted from the target. L represents the maximum number of shifts (both left and right) used in the conversion. The RALUT is accessed $2^R \cdot (L \cdot 2 + 1) + 1$ times for each conversion, which increases the latency of the serial implementation (or the area of the parallel implementation). We generally use this method for generating offline MDLNS representations.

6.5.5 Comparison of Binary-to-Two-Digit 2DLNS Conversion Methods

In order to compare the above methods using MRRE and ARRE, we select an integer range of conversion between 0 and 256, with $D = 3$, $B = 5$, and $R = 3$. The results can be found in Table 6.8.

Although the MRRE slightly improves from method to method, a lower ARRE is more favorable, as it increases the overall accuracy of the representation. We can also see that the performance of the extended-brute-force method tapers off for $L > 1$, indicating that the extra hardware probably is not worth the improved accuracy.

The improvement of the approximations from method to method is generally consistent for any good choice for D. For example, if $D = 4/3$, the MRRE and ARRE are lower than when $D = 3$, as seen in Table 6.9.

TABLE 6.8

MRRE and ARRE for Integers between 0 and 256 for $D = 3$ and $R = 3$

Method	Quick	High/Low	Brute Force	Extended Brute Force $L = 1$	Extended Brute Force $L = 2$
MRRE	2.84900E-03	2.84900E-03	2.32450E-03	1.65529E-03	1.65529E-03
ARRE	4.20419E-04	2.57948E-04	1.02366E-04	8.88704E-05	8.84556E-05

TABLE 6.9

MRRE and ARRE for Integers between 0 and 256 for $D = 4/3$ and $R = 3$

Method	Quick	High/Low	Brute Force	Extended Brute Force $L = 1$	Extended Brute Force $L = 2$
MRRE	2.32450E-03	2.03339E-03	2.32450E-03	1.65529E-03	1.65529E-03
ARRE	2.74714E-04	2.28989E-04	8.27815E-05	6.81324E-05	6.81324E-05

6.6 Multidigit 2DLNS Representation ($n > 2$)

The methods used in single-digit and two-digit 2DLNS conversion can be extended to operate with n-digit 2DLNS with $n > 2$. All methods are briefly reviewed, below, for completeness, although some methods are not very practical.

6.6.1 Multidigit 2DLNS-to-Binary Conversion

Just as single-digit 2DLNS-to-binary conversion architectures can be easily extended to two-digit systems, they may also be extended to n-digit systems. The serial implementations require few changes, but result in decreased throughput. Conversely, the parallel system requires more area for additional conversion LUTs, but there is no performance degradation—only an increase in latency.

6.6.2 Binary-to-Multidigit 2DLNS Conversion (Quick Method)

The quick binary-to-two-digit 2DLNS conversion process can be easily extended into an n-digit system. The serial implementation of the n-digit conversion requires extra states in the controlling state machine to produce approximations for the remaining digits. The output latency is a multiple of n, but the throughput rate decreases. The parallel implementation requires additional pipeline stages (normalizers, RALUTs, etc.) to meet the n-digit specifications. The latency is n cycles, but maintains its operating throughput rate.

6.6.3 Binary-to-Multidigit 2DLNS Conversion (High/Low Method)

The high/low method of conversion does not scale as well as the quick method since there are many possibilities to consider when converting from binary into 2DLNS. For the serial implementation a stack-based state machine is used to traverse all the possible representations for the n-digit 2DLNS representation in order to find the best representation. The parallel implementation requires double the number of conversion components for each pipeline stage from the previous stage (i.e., 1, 2, 4, 8, 16, …). In total, $2^n - 1$ conversion components will be needed. Just as in the quick method, the latency will increase, but the throughput rate will remain the same. The final stage will require $2^{n-1} - 1$ comparators in order to determine the best n-digit 2DLNS approximation (see Figure 6.9).

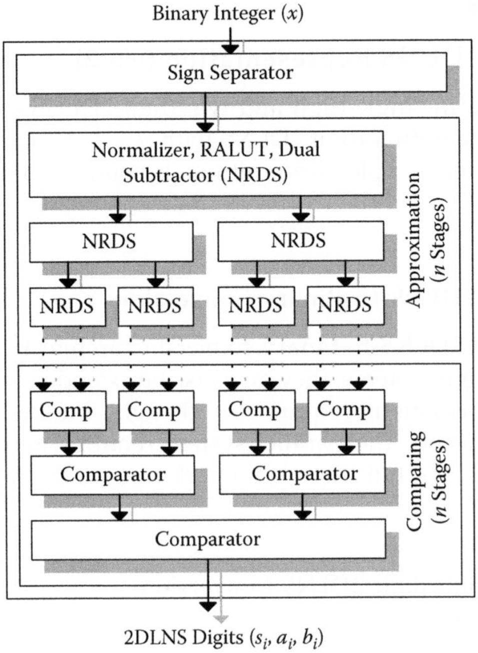

FIGURE 6.9
High/low binary-to-n-digit 2DLNS parallel converter structure. (From R. Muscedere, V. Dimitrov, G.A. Jullien, W.C. Miller, Efficient Techniques for Binary-to-Multidigit Multidimensional Logarithmic Number System Conversion Using Range-Addressable Look-Up Tables, *IEEE Transactions on Computers*, 54, 257–271, 2005. Copyright © 2005 IEEE. With permission.)

6.6.4 Binary-to-Multidigit 2DLNS Conversion (Brute-Force Method)

The brute-force method can be implemented for n-digit systems; however, the number of accesses to the RALUT grows exponentially with n to $(2^R \cdot (L \cdot 2 + 1) + 1)^{n-1}$. This is because all the first $n - 1$ digit combinations are generated and the last digit, the nth, minimizes the final error. Although practical for the two-digit case, it is much less desirable for more than two digits. While it is possible to implement n-digit brute-force conversion circuits, it is recommended to simply increase R in order to improve the 2DLNS resolution.

6.7 Extending to More Bases

The conversion methods presented can be easily extended into MDLNS representations with more than two bases (assuming one of the bases is still 2). Since all the methods presented in this chapter rely only on information

TABLE 6.10

MDLNS-to-Binary Conversion LUT for $D_1 = 3$, $D_2 = 5$, and $R = 2$

Input		Output	
b_1	b_2	$M(b_1, b_2)$ (Combined Mantissas)	$e(b_1, b_2)$ (Combined Exponent)
0	0	1.0000000000	0
–1	–1	1.0666666667	–4
–2	1	1.1111111111	–1
–2	–2	1.1377777778	–8
1	–1	1.2000000000	–1
0	1	1.2500000000	2
0	–2	1.2800000000	–5
–1	0	1.3333333333	–2
–2	–1	1.4222222222	–6
1	0	1.5000000000	1
0	–1	1.6000000000	–3
–1	1	1.6666666667	0
–1	–2	1.7066666667	–7
–2	0	1.7777777778	–4
1	1	1.8750000000	3
1	–2	1.9200000000	–4

about the mantissa and exponent, multiple bases can be merged, generating a single mantissa and exponent. For example, Table 6.10 shows a MDLNS-to-binary conversion LUT using bases 2, 3, and 5, and a word length of 2 bits for the indices using bases 3 and 5. The non-base 2 indices can be combined through simple word concatenation to generate a single word that can address the table to determine the proper mantissa and exponent.

The same method can be used to generate a RALUT for binary-to-MDLNS conversion. Similarly, the combined second base RALUT output can be easily separated to find the non-base 2 indices to generate the MDLNS representation.

The next chapter on MDLNS addition and subtraction uses three bases, 2, 3, and 5, which illustrates this extension effectively.

6.8 Physical Implementation

A number of designs in the literature [6–8] make use of the two-digit 2DLNS high/low serial converter as it is a compact design. One in particular [7] is an 8-band MDLNS filterbank design for use in a digital hearing instrument (fabricated in a TSMC 0.18 μm CMOS process). The input binary data are in a 16-bit two's complement format and are converted to 2DLNS with the second base exponents limited from –14 to 14, so that overflow never occurs when the input data are multiplied with the coefficients (the second base exponent has a range from –2 to 1). The second base is 1.28308348549366, optimally chosen, using an exhaustive search technique (see Chapter 8), over all the filterbank coefficients, to minimize the filter response error from the ideal design. Any base can be used since the conversion process only performs pattern matching on the generated mantissas of the base. Since the filterbank is intended for processing speech with a required very low power operation, the serial implementation is ideal as it minimizes both power and area. The design core is 1 mm × 1 mm and 1.67 mm × 1.67 mm, including I/O pads. Half of the core area is occupied by the converter because the RALUT is built with standard cells (as in Figure 6.3). A custom RALUT generator tool has since been created that uses modern dynamic address decode techniques to reduce power, interconnect, and area. A more detailed account can be found in Chapter 9.

6.9 Very Large-Bit Word Binary-to-DBNS Converter

There are applications, particularly in cryptography systems, where very large binary numbers, of the order of hundreds of bits, are used. Such converted binary numbers will have many more DBNS digits than considered in earlier sections of this chapter. Since an exact DBNS representation of the original number is required for these applications, the DBNS exponents are limited to being whole numbers, hence eliminating any possibility of fractional digits. We therefore refine our ranges of a_i and b_i to $a_i = \{0, \ldots, 2^B - 1\}$ and $b_i = \{0, \ldots, T\}$; b_i is no longer limited to use the fully available bit width, R. There are some approaches in literature that use DBNS arithmetic to minimize the overall number of calculations for different encryption algorithms [9,10]. Their effective performance characteristics are all based on the number of digits in the final representation. Ideally we would like this number to be small; however, there should be a balance between hardware and the resulting number of digits.

For the application of binary-to-DBNS conversion, the RALUT's comparators hold the normalized representation of all single-digit DBNS values, which is only the 3^{b_i} component, as the normalization function removes the 2^{a_i} component. We can therefore expect there to be $T + 1$ rows in the table. For DSP or low-bit word applications, the width of the address decode, M, would typically be one bit larger than the input word, as it is not necessary to hold all the bits of each case of 3^{b_i}, only enough to achieve an accuracy of +/– half a bit from the original value. This is an acceptable balance for hardware as the normally infinite binary representations, 1/3, for example, would be truncated and rounded toward positive infinity.

6.9.1 Conversion Methods for Large Binary Numbers

As the DBNS is a redundant number system, this allows us the flexibility of creating different algorithms for generating the series of DBNS digits. For encryption applications, the sign of each digit (s_i) may be limited to 1 or both 1 and –1, depending on the particular encryption algorithm. We have already introduced the greedy algorithm and some variations. We use them, as below, but with specific limitations to be developed for large binary numbers.

Algorithm 6.4: A Greedy Algorithm for DBNS with

Input a positive integer x.
Output the sequence of exponents (a_n, b_n) leading to one DBNS representation of x.

Step 1: While $x > 0$ do
Step 2: Find $z = 2^a \times 3^b$, the largest DBNS representation less than or equal to x
Step 3: Save a and b
Step 4: $x = x - z$

Algorithm 6.5: A Modified Greedy Algorithm for DBNS with

Input a positive integer x.
Output the sequence of exponents (s_n, a_n, b_n) leading to one DBNS representation of x.

Step 1: $s = 1$
Step 2: While $x > 0$ do
Step 3: Find $z = 2^a \times 3^b$, the closest DBNS representation to x
Step 4: Save s, a, and b
Step 5: If $x < z$; $s = -s$
Step 6: $x = |x - z|$

For very large input words (more than 24 bits) for use in encryption applications, we can use these algorithms, but we must constrain the system such that there should be no truncation of the digits, as it could result in an inaccurate DBNS representation. It is this need for accuracy that results in large values of M, which translate to potentially large tables. For example, if $T = 7$ each address decode would require at most $\log(3^7)/\log(2)$ bits, or 12 single-bit comparators ($8 \times 12 = 96$ for the whole RALUT), an acceptable number for implementation. However, if $T = 31$ each address decode would require $\log(3^{31})/\log(2) = 50$ bit words or $32 \times 50 = 1600$ total comparators, a much larger system and possibly unacceptable for implementation. To maintain accuracy we must set $M = T \cdot \log(3)/\log(2)$. Although we have limited the exponents to be only positive, the greedy methods may still generate fractional digits by selecting overflow or negative values of a_i. To eliminate this possibility we include additional functionality in the RALUT address decoder, to select rows that meet an exponent range as well as the address ranges.

We demonstrate the hardware procedure of the first method, where $s_i = 1$ and $T = 3$. The RALUT is configured to match on addresses smaller than that in the table and exponents greater than that in the table (see Table 6.11). Since the first bit of all the addresses in the RALUT is 1, we can remove them in a practical implementation; however, we will not do so here for completeness of this example. Our goal is to convert the binary value $10111_{(2)}$ (23) to a DBNS representation. The procedure consists of a repeating process of three steps: normalization, table lookup, and subtraction. In this example, we obtain a three-digit result of (1,2), (2.0), (0,0) = 18 + 4 + 1 = 23.

The RALUT is somewhat different for the modified greedy method as we use the address comparators to select the closest DBNS digit to the normalized value (see Table 6.12). Here we set the matching address range to the midpoint between those of the previous example. Because of this we have to

TABLE 6.11

RALUT Contents for the Greedy Method

Input		Output		
Address	**Exponent**	**Mantissa**	**Exponent**	***b***
10000	000	10000	000	00
10010	011	10010	011	10
11000	001	11000	001	01
11011	100	11011	100	11

TABLE 6.12

RALUT Contents for the Modified Greedy Method

Input		Output		
Address	**Exponent**	**Mantissa**	**Exponent**	***b***
10000	000	010000	000	00
10001	011	010010	011	10
10101	001	011000	001	01
11010	100	011011	100	11
11110	001	100000	001	00

include an additional row to compensate for the cyclical nature of the representation. In this case of converting 23 to a DBNS representation the final result is a two-digit representation of (1,3,1), (–1,0,0) = 24 – 1 = 23. Clearly the inclusion of negative digits can result in fewer overall digits compared to the previous method, but this negative representation can not necessarily be used in all applications.

6.9.2 Reducing the Address Decode Complexity

As the complexity of the RALUT depends mostly on the number of bits in the address decode, it is natural to try to reduce them. We can use the same technique as in the DSP converter case by truncating the least significant digit and rounding toward positive infinity. We do not truncate the values in the table output, as they must remain accurate to maintain no error in the representation. In the case of Table 6.11 (greedy method), we can do this for up to two bits before the addresses are no longer unique. If we go through the hardware method again with the reduced input addresses we will arrive at the same representation for the case of 23. We expect there to be more, on average, digits in each representation, as we've changed the matching characteristics of the RALUT, but this penalty is balanced by the removal of these least significant bits. We can do the same for Table 6.12 (modified greedy method). In this particular case we can remove only one bit before the addresses are no longer unique.

6.9.3 Results and Discussion

The above example is purely illustrative and not necessarily practical for arriving at any results. To demonstrate the benefit of dropping the least significant address bits, we've generated 100,000 unique and uniformly random 593-bit prime numbers (used on GF(2^{593}) arithmetic, but can be applied to any number of bits) and tested the effects on the two methods for various values of T by removing the least significant address bits until any of the representations changed. We also made note of how many bits could be dropped before the average number of digits in the representation increased by only 1. The full results can be seen in Table 6.13 for the greedy method and Table 6.14 for the modified greedy method.

For values of T where the number of comparators originally exceeds 1,000, the savings is over 79% and 85% for the greedy and modified greedy methods, respectively. We can achieve a higher reduction but at the cost of additional digits. In the case of the greedy method the number of digits continues to slowly increase as we reach the maximum address bits removable, whereas in the modified greedy case, the digits increase immediately as we remove the maximum number of address bits. In fact, with three of the T's selected (63, 127, and 374), the maximum number of bits can be dropped without exceeding one additional digit, on average. Our results also indicate

TABLE 6.13

Comparator Reduction Results for the Greedy Method

	Before Reduction		Reduction before Any Additional Digits				Reduction with One Additional Digit				Maximum Reduction			
T	C/R	TC	CR	TC	%S	AVG	CR	TC	%S	AVG	CR	TC	%S	AVG
7	12	88	0	88	0.0	122.1	3	64	27.3	122.9	6	40	54.5	129.8
15	24	368	1	352	4.3	102.9	14	144	60.9	103.7	18	80	78.3	115.7
31	50	1,568	23	832	46.9	89.3	39	320	79.6	90.3	43	192	87.8	97.7
63	100	6,336	71	1,792	71.7	79.2	88	704	88.9	80.0	90	576	90.9	81.6
127	202	25,728	173	3,584	86.1	74.2	189	1,536	94.0	74.8	192	1,152	95.5	77.6
255	405	103,424	375	7,424	92.8	69.6	392	3,072	97.0	70.4	395	2,304	97.8	73.7
374	593	222,000	563	10,875	95.1	68.9	580	4,500	98.0	69.8	582	3,750	98.3	71.5

Note: C/R = comparators per row, TC = total comparators, CR = comparators removed, %S = percent savings, AVG = average digits per representation.

TABLE 6.14

Comparator Reduction Results for the Modified Greedy Method

	Before Reduction		Reduction before Any Additional Digits				Reduction with One Additional Digit				Maximum Reduction			
T	C/R	TC	CR	TC	%S	AVG	CR	TC	%S	AVG	CR	TC	%S	AVG
7	12	99	1	90	9.1	101.9	5	54	45.5	102.5	6	45	54.5	105.0
15	24	391	0	391	0.0	88.2	16	119	69.6	88.6	18	85	78.3	93.8
31	50	1,617	21	924	42.9	78.0	42	231	85.7	78.6	43	198	87.8	81.7
63	100	6,435	73	1,690	73.7	70.1	90	585	90.9	70.2	90	585	90.9	70.2
127	202	25,929	170	3,999	84.6	66.1	192	1,161	95.5	66.8	192	1,161	95.5	66.8
255	405	103,828	375	7,453	92.8	62.4	394	2,570	97.5	62.7	395	2,313	97.8	63.5
374	593	222,592	563	10,904	95.1	61.8	582	3,760	98.3	62.1	582	3,760	98.3	62.1

Note: C/R = comparators per row, TC = total comparators, CR = comparators removed, %S = percent savings, AVG = average digits per representation.

that it may not be necessary to use a very large value of T to minimize the number of digits in the DBNS representation, as they appear to begin to plateau after $T = 127$.

6.10 Summary

A framework for converting binary-to-single-digit 2DLNS has been introduced by analyzing and reversing the process of conversion from single-digit 2DLNS to a binary representation. To improve hardware implementation scalability, a RALUT, a novel memory device with a range addressing system, has been introduced. Using this efficient binary-to-single-digit 2DLNS converter, three types of multidigit 2DLNS conversion methods have been developed: quick, high/low, and brute force. Each of these methods offers either simpler implementation or more accurate 2DLNS approximations. Furthermore, these converters may also be extended to handle multiple-base MDLNS representations. The two-digit serial converter design has been implemented in a number of designs, even though it is completely realized with standard cells. We expect to be able to lower the power consumption and area by using low-power library cells and a custom RALUT implementation. All of these methods have been implemented in fully parameterized Verilog HDL code.

We have also shown that the binary-to-DBNS converter concept, originally designed for low-bit words, a fixed number of DBNS digits, and minimum error, can be scaled to handle large-bit words, variable digits, and no representation error with a minimal change in address decode complexity. This large binary word extension to the converter theory is targeted to cryptography applications.

All software, source files, and detailed results can be found at http://research.muscedere.com.

References

1. R. Muscedere, V. Dimitrov, G.A. Jullien, W.C. Miller, Efficient Techniques for Binary-to-Multidigit Multidimensional Logarithmic Number System Conversion Using Range-Addressable Look-Up Tables, *IEEE Transactions on Computers*, 54, 257–271, 2005.
2. R. Muscedere, A Hardware Efficient Very Large Bit Word Binary to Double Base Number System Converter for Encryption Applications, in *Proceedings of IEEE International Symposium on Circuits and Systems*, ISCAS 2007, 2007, pp. 1373–1376.

3. V.S. Dimitrov, G.A. Jullien, W.C. Miller, Theory and Applications of the Double-Base Number System, *IEEE Transactions on Computers*, 48, 1098–1106, 1999.
4. V.S. Dimitrov, J. Eskritt, L. Imbert, G.A. Jullien, W.C. Miller, The Use of the Multi-Dimensional Logarithmic Number System in DSP Applications, in *Proceedings of 15th IEEE Symposium on Computer Arithmetic*, 2001, pp. 247–254.
5. I. Koren, *Computer Arithmetic Algorithms*, Prentice-Hall, Englewood Cliffs, NJ, 1993.
6. M. Azarmehr, R. Muscedere, A RISC Architecture for 2DLNS-Based Signal Processing, *International Journal of High Performance Systems Architecture*, 3, 149–156, 2011.
7. H. Li, G.A. Jullien, V.S. Dimitrov, M. Ahmadi, and W. Miller, A 2-Digit Multidimensional Logarithmic Number System Filterbank for a Digital Hearing Aid Architecture, in *Proceedings on IEEE International Symposium on Circuits and Systems*, ISCAS 2002, 2002, vol. 2, pp. II-760–II-763.
8. R. Muscedere, V. Dimitrov, G. Jullien, W. Miller, A Low-Power Two-Digit Multi-Dimensional Logarithmic Number System Filterbank Architecture for a Digital Hearing Aid, *Eurasip Journal on Applied Signal Processing*, 2005, 3015–3025, 2005.
9. V. Dimitrov, L. Imbert, A. Zakaluzny, Sublinear Constant Multiplication Algorithms, in *Advanced Signal Processing Algorithms, Architectures, and Implementations XVI*, San Diego, CA, 2006, pp. 631305–631309.
10. V.S. Dimitrov, L. Imbert, P.K. Mishra, *Fast Elliptic Curve Point Multiplication Using Double-Base Chains*, in Cryptology ePrint Archive, Report 2005/069, 2005.

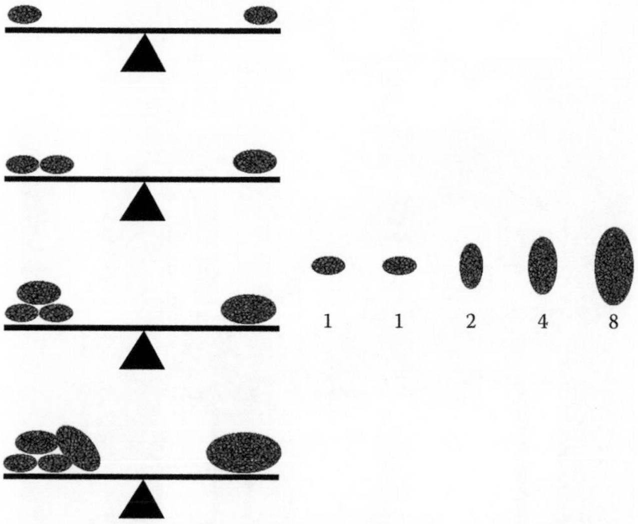

FIGURE 1.1
Four-thousand-year-old binary number system.

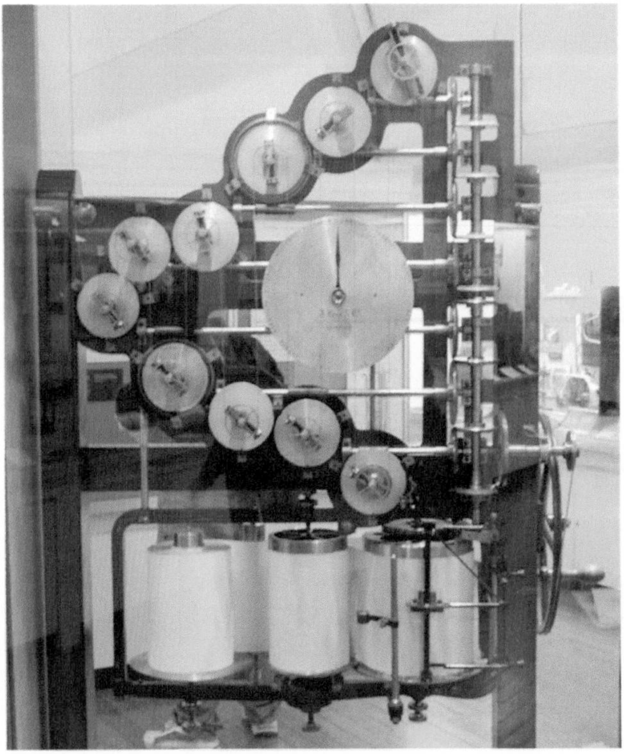

FIGURE 1.2
Lord Kelvin's tide predicting machine (1876), Science Museum, London. (Copyright 2005 William M. Connolley under the GNU Free Documentation License.)

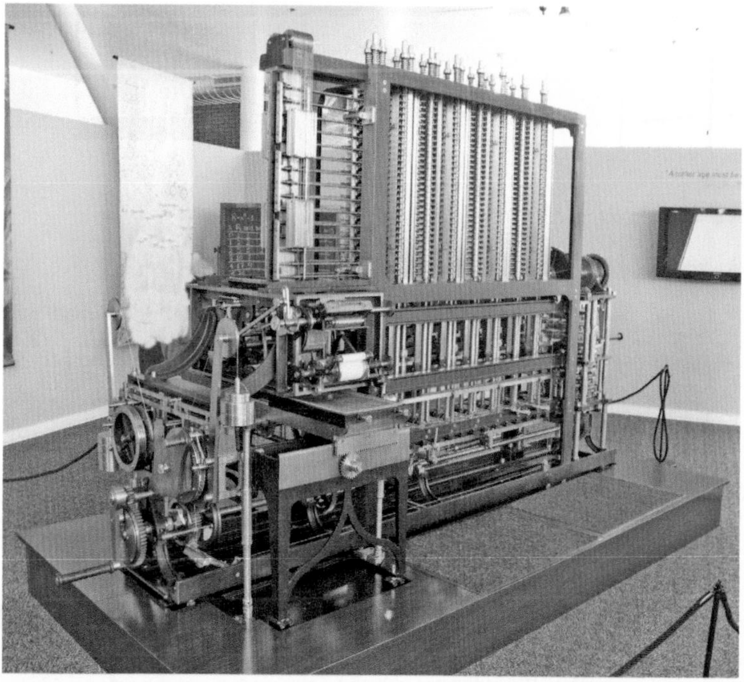

FIGURE 1.3
Babbage's difference engine, Computer History Museum, Mountain View, California. (Copyright 2009 Allan J. Cronin under the GNU Free Documentation License.)

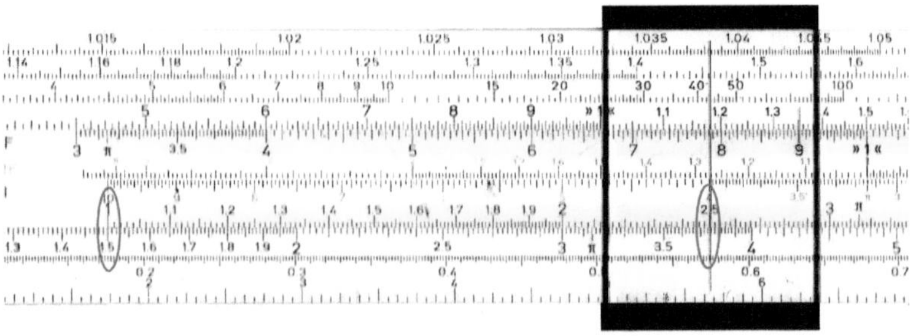

FIGURE 1.4
Multiplication of 1.5 by 2.5 on a slide rule.

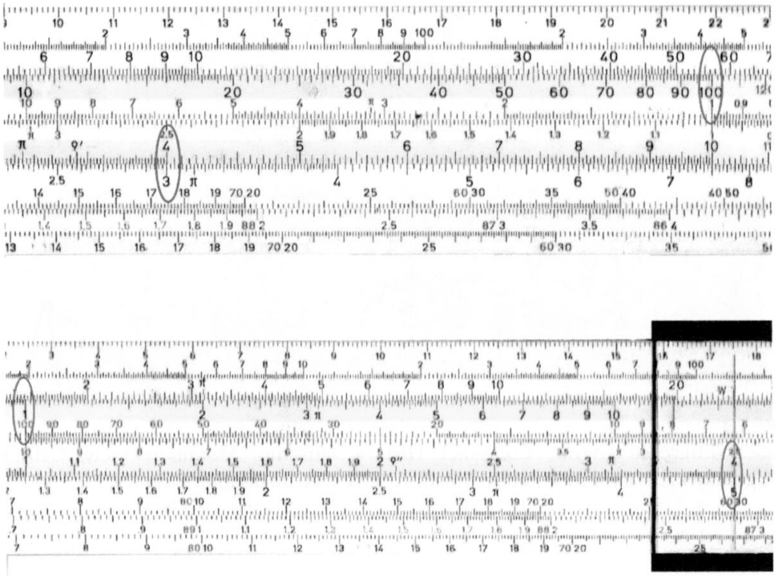

FIGURE 1.5
Three moves to compute the slide rule trick.

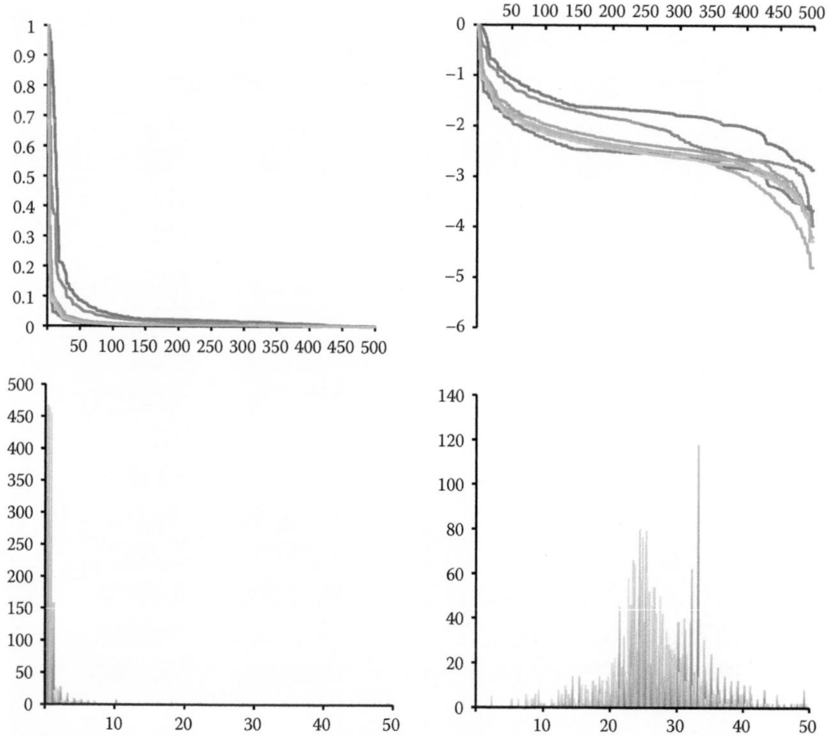

FIGURE 2.16
Coefficient statistics for eight different FIR filter designs. Number representations: linear (left) and logarithmic (right). Upper plots show the ordered coefficient absolute values, lower plots show 50-bin histograms of the dynamic range usage.

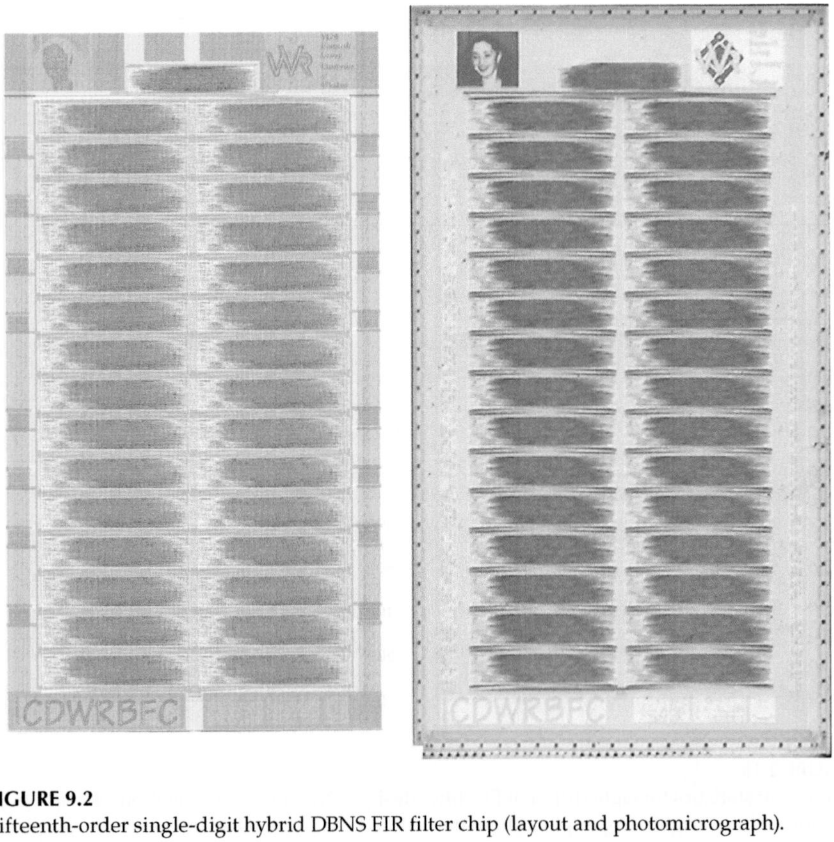

FIGURE 9.2
Fifteenth-order single-digit hybrid DBNS FIR filter chip (layout and photomicrograph).

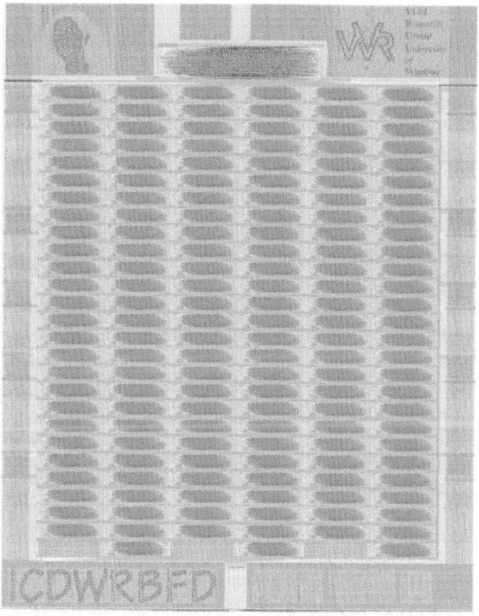

FIGURE 9.3
Fifty-third-order two-digit DBNS FIR filter (layout only).

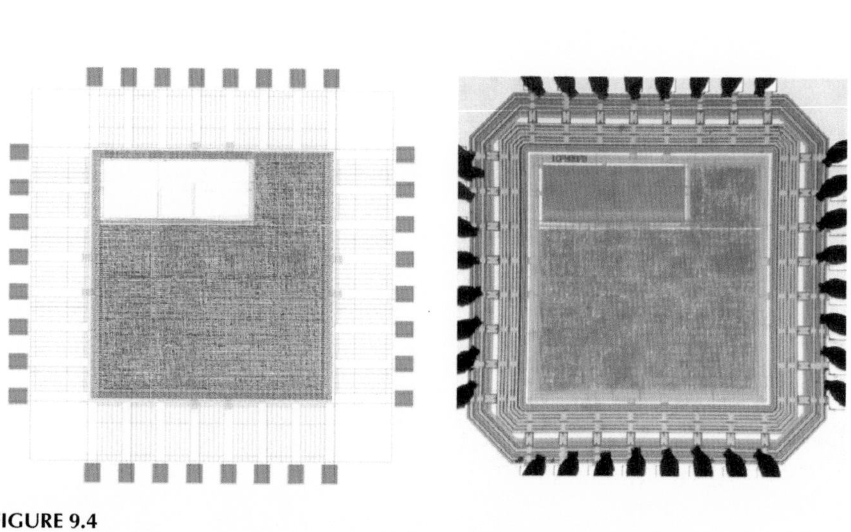

FIGURE 9.4
Seventy-third-order low-power two-digit MDLNS eight-channel filterbank (layout and photomicrograph). (From H. Li, G.A. Jullien, V.S. Dimitrov, M. Ahmadi, W. Miller, A 2-Digit Multidimensional Logarithmic Number System Filterbank for a Digital Hearing Aid Architecture, in *IEEE International Symposium on Circuits and Systems,* ISCAS 2002, 2002, vol. 2, pp. II-760–II-763. Copyright © 2002 IEEE.)

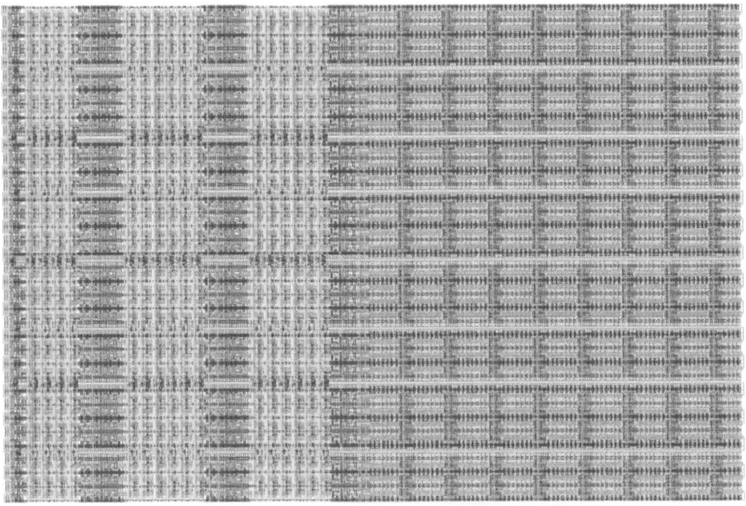

FIGURE 9.13
Screen dump of a 29 × 16 × 52 RALUT (0.18 μm TSMC [Taiwan Semiconductor Manufacturing Company] process). (From R. Muscedere, K. Leboeuf, A Dynamic Address Decode Circuit for Implementing Range Addressable Look-Up Tables, in *IEEE International Symposium on Circuits and Systems*, ISCAS 2008, 2008, pp. 3326–3329. Copyright © 2008 IEEE.)

7

Multidimensional Logarithmic Number System Addition and Subtraction

7.1 Introduction

The two-dimensional logarithmic number system (2DLNS) systems, already covered in previous chapters, utilize architectures that favor the easy operations of multiplication and division, but try to minimize any use of addition or subtraction, as they are costly and difficult functions requiring large lookup tables (LUTs) whose sizes depend exponentially on the bit widths of the 2DLNS exponents. This problem is not restricted to the multidimensional logarithmic number system (MDLNS), but also with the logarithmic number system (LNS) [1,2]. One of the most popular MDLNS architectures is the inner-product half-domain processor, which performs multiplication in the MDLNS domain and accumulates in the binary domain, therefore eliminating any further opportunity to process in the MDLNS domain. MDLNS processing can only continue if either the binary output is converted back to MDLNS (via a binary-to-MDLNS converter, covered in Chapter 6) or a native MDLNS addition/subtraction operation is used. No such circuit has been published to date, and it is assumed that the implementation would consist of mainly LUTs. Such a circuit would allow designers to explore MDLNS architectures for other systems and applications.

This chapter introduces a realizable method to perform single-digit MDLNS addition and subtraction using very small LUTs and range-addressable lookup tables (RALUTs), through the use of a single-base domain. The fundamental single-digit addition/subtraction operations can be extended to allow the use of multidigit MDLNS and to improve quantization accumulation error (for systems targeted to digital signal processing (DSP) applications).

7.2 MDLNS Representation

For the LNS, the logarithmic representation, a, of the number, x, is given by the relationship in Equation (7.1), where s is the sign of the number (either –1, 0, or + 1) and the base r is usually 2 to simplify the hardware implementation in binary (digital) technologies.

$$x = s \cdot r^a \tag{7.1}$$

Although 2 is often selected as the value for r, there has been some research into using other bases to reduce power by minimizing switching [3]; however, there has been no discussion on the implementation challenges. The extra degrees of freedom obtained from using MDLNS representations introduce new complexities in the addition and subtraction process. The LNS addition and subtraction process is simplified due to the monotonic relationship between x and a. Unfortunately this solution is not applicable to the MDLNS since there is no monotonic relationship between x and the exponents on the multiple digits and bases (similar issues as in binary-to-MDLNS conversion). Since MDLNS is a generalization of LNS, current LNS solutions for addition and subtraction may apply to only a few specific instances of MDLNS. The intention of this chapter is to describe an accurate generalized solution.

7.2.1 Simplified MDLNS Representation

To simplify the presentation of the proposed processes we will restrict ourselves to a single-digit MDLNS where the representation of an input, x, is shown in Equation (7.2):

$$x = s \cdot \prod_{j=1}^{k} D_j^{b_j} \tag{7.2}$$

where D_1 (sometimes referred to as the binary base) is selected to be 2 (as in LNS to significantly simplify hardware implementation). The remaining bases, $D_2, D_3, \ldots, D_k$, should not be multiples of other bases and do not necessarily have to be integers (see Chapter 8), as their value bears no impact on the complexity of the hardware implementation. Since we intend to implement MDLNS in hardware, we limit the range of the exponents such that $b_j = \{L_j, \ldots, H_j\}$. We note, however, that the binary base and its exponents simply translate to bit shifts. We therefore measure the complexity of the system by the number of combinations of x (independent of s) when $j > 1$. This complexity, R, shown in Equation (7.3), is equivalent to the number of table entries or rows used for conversion both to and from binary (see Chapter 6).

$$R = \prod_{j=2}^{k} (H_j - L_j + 1) \tag{7.3}$$

The range of b_1 dictates the range of the representation, while $\{b_2, b_3, \ldots, b_k\}$ dictates the resolution. To simplify the comparison of the existing and proposed methods for addition and subtraction in the MDLNS, we will set C_1 and $\{C_2, C_3, \ldots, C_k\}$ to the number of bits required to represent b_1 and $\{b_2, b_3, \ldots, b_k\}$, respectively.

7.3 Simple Single-Digit MDLNS Addition and Subtraction

The simplest method to perform addition and subtraction in MDLNS is to use a pure LUT scheme. The minimum size in bits of each single-digit MDLNS operand is given by Equation (7.4):

$$A = 1 + \sum_{j=1}^{k} \text{ceiling}(\log_2 (H_j - L_j + 1)) \tag{7.4}$$

or in terms of C:

$$A = 1 + C_1 + C_{2\to k} \tag{7.5}$$

The sign is reduced to a single bit as cases of $s = 0$ can be handled with logic alone. The sign and exponents of each of the two operands are combined into a single word and used as the address of a LUT. The output of the LUT contains the combined sign and exponents of the best MDLNS representations for the addition and subtraction operations. Although simple and quite fast, the hardware implementation complexity is based solely on this combined bit width, which also includes information on the binary base. Assuming only two operands, the number of input and output bits to the combined LUT will be $2A$. For example, if $k = 3$, $D_j = 2, 3, 5$, and $b_j = \{-16, \ldots, 15\}, \{-1, 0, 1\}, \{-1, 0, 1\}$, then $C_1 = 5$, $C_{2\to k} = 4$, resulting in $A = 10$, which generates a LUT with 20 bits for the input and output (20 Mbit). Should the constraints of the system be changed such that either C increases by only 1 bit, the resulting table would be 88 Mbit, which is over four times the original. Because of the exponential growth with input address, this method is generally only acceptable for systems where A is quite small.

7.4 Classical Method

A more efficient method of addition/subtraction in logarithms with base D is to multiply one of the addends (D^x) by a factor D^z (for addition, see Equation (7.6)) or D^w (for subtraction, see Equation (7.7)). In terms of logarithms, z or w is added to x, as given in Equations (7.6) and (7.7).

$$\begin{aligned} D^x \cdot D^z &= D^x + D^y \\ D^z &= 1 + D^{y-x} \\ z &= \log_D(1 + D^{y-x}) \end{aligned} \tag{7.6}$$

$$\begin{aligned} D^x \cdot D^w &= D^x - D^y \\ D^w &= 1 - D^{y-x} \\ w &= \log_D(1 - D^{y-x}) \end{aligned} \tag{7.7}$$

This method was originally described by Leonelli in 1803 and was widely used in the 19th century after Gauss popularized it. The typical relationships between the input and output to the multiplication factors, obtained from Equations (7.6) and (7.7), are shown in figures 7.1 and 7.2, respectively.

In both of these figures, we see that as the difference $y - x$ exceeds 0, the factor approaches this difference. We only consider the left quadrant of these graphs (i.e., $x > y$), as their limits are finite and their values are generally rather small. This is desirable as MDLNS/LNS can represent small numbers near 1 with good accuracy.

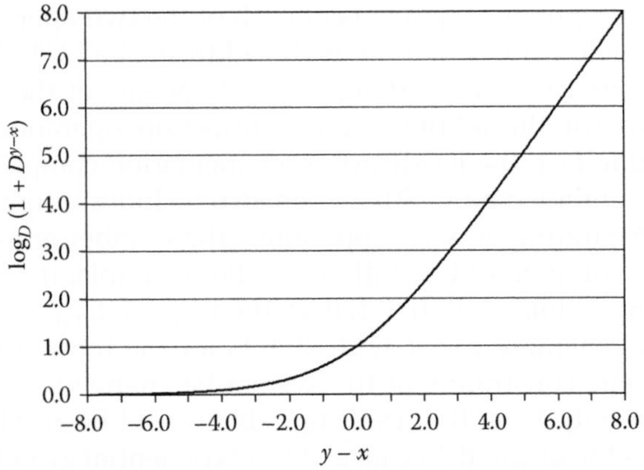

FIGURE 7.1
LNS addition relationship for $D = 2$.

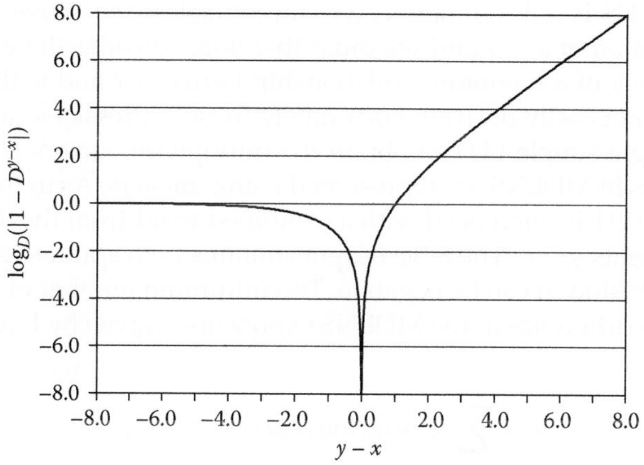

FIGURE 7.2
LNS subtraction relationship for $D = 2$.

7.4.1 LNS Implementation

Addition and subtraction operations in the LNS have been implemented in hardware using many different number representations [4–8] (integer, fixed-point, floating point, and integer rational numbers). Depending on the method of implementation, D^z and D^w may be either extracted from a LUT (addressed by $y - x$), calculated exactly from the above expressions, or interpolated from smaller tables. The system constraints and accuracy required will dictate the implementation method used.

7.4.2 MDLNS Implementation

This classical method can be extended to operate in a single-digit MDLNS by adding multiple bases (see Section 7.2) to Equations (7.6) and (7.7).

$$\begin{gathered}\prod_{i=1}^{k} D_i^{x_i} \cdot \prod_{i=1}^{k} D_i^{z_i} = \prod_{i=1}^{k} D_i^{x_i} + \prod_{i=1}^{k} D_i^{y_i} \\ \prod_{i=1}^{k} D_i^{z_i} = 1 + \prod_{i=1}^{k} D_i^{y_i - x_i}\end{gathered} \tag{7.8}$$

$$\begin{gathered}\prod_{i=1}^{k} D_i^{x_i} \cdot \prod_{i=1}^{k} D_i^{w_i} = \prod_{i=1}^{k} D_i^{x_i} - \prod_{i=1}^{k} D_i^{y_i} \\ \prod_{i=1}^{k} D_i^{w_i} = 1 - \prod_{i=1}^{k} D_i^{y_i - x_i}\end{gathered} \tag{7.9}$$

Since MDLNS is not monotonic, we cannot isolate the above expressions for the condition of $x > y$ and we must therefore consider the entire range. Due to the lack of a monotonic relationship between x and y, the values of z_i and w_i are not easily derived. Fortunately these expressions can be implemented using a simple LUT to obtain the multiplying factors D^z and D^w as the exponents in MDLNS are represented using integers. As in the pure LUT method, the LUT is referenced with a combined word from the difference of all the exponents $y_i - x_i$. The table output contains both z_i and w_i along with a sign bit for D^w since it can be negative. The minimum number of bits for representing the difference in the MDLNS exponents is given by Equation (7.10):

$$B = \sum_{j=1}^{k} \text{ceiling}\left(\log_2\left(2H_j - 2L_j + 1\right)\right) \tag{7.10}$$

or in terms of C:

$$B = C_1 + C_{2\to k} + k \tag{7.11}$$

We do not need to consider the input signs in the table as additional logic can determine the final sign and which operation to perform. Using this classical method and assuming only two operands, the number of input and output bits to the LUT will be B and $2B + 1$, respectively. Using the previous example, $B = 12$, the input and output bits to the LUT are 12 and 25, resulting in a 100 kbit LUT. Again, should the constraints of the system be changed such that either C increases by only 1 bit, the resulting table would be 216 kbit, which is just over two times the original. For this example, the table for the classical method is at least 10 times smaller than that for the exact LUT method, and in general, it scales better with changes in B. However, this reduction in table space comes at a cost in terms of accuracy. When D^z or D^w is multiplied by D^x, the result could generate an MDLNS overflow, a situation where one or more of the exponents exceeds the intended range $(\{L_j, \ldots, H_j\})$. An overflow is usually handled by a compensation circuit (see Chapter 5), which uses a unity approximant to recover from the overflow (see Table 7.1 for an example). Of course, this recovery operation results in an error in the final computed output. As it is likely that the approximations of D^z and D^w will not be error-free, an additional bit for each output is used to specify the preferred direction of the overflow correction, either above or below 1. This will result in a more accurate output. The number of table output bits is therefore $2B + 3$. Using the previous example where $B = 12$, the table inputs and outputs would be 12 and 27 bits or a 108 kbit LUT. This additional bit will not be a detriment to the table scalability. In comparison to the exact LUT method, the classical method offers considerable areas savings but at the cost of accuracy since the approximations of D^z and D^w are

TABLE 7.1

Unity Approximants for $k = 3$, $D_j = 2, 3, 5$, and $a_j = \{-16, \ldots, 15\}, \{-1, 0, 1\}, \{1, 0, 1\}$

a_1	a_2	a_3	$\prod_{j=1}^{k} D_j^{a_j}$
3	−2	0	0.888889
−4	1	1	0.937500
3	1	−2	0.960000
−3	−1	2	1.041667
4	−1	−1	1.066667
−3	2	0	1.125000

not likely error-free and there is a high probability of MDLNS overflow. This may be acceptable in some applications considering the trade-off in silicon area saved.

If we carefully examine the contents of the example addition LUT, only 140 of the 2,048 entries contain valid data (for the condition of $x > y$). The remainder include factor representations for adding large numbers to small numbers, cases in which the sum is either one of the addends. These 140 entries are made up of 25 unique values that represent the factors between 1 and 2. It would appear that the classical method is not very efficient at encoding the output, as there is a considerable amount of duplication. These 25 of the 2,048 entries are the only values required for computing MDLNS addition. The sparseness of this table is attributed to the multiple bases in MDLNS, an undesirable trait that makes direct table reduction difficult.

From the LNS expressions for the classical method, there is a direct monotonic relationship between the function inputs ($y - x$) and outputs (as seen in Figures 7.1 and 7.2). From this, the range of usable table data can be determined and all other data can be ignored or eliminated. Due to the predictable trends of the expression, circuits are often developed to generate similar results to that of a table at a smaller cost in area. Unfortunately, this comes at the cost of accuracy. If a similar relationship existed in MDLNS, additional enhancements could be made to reduce the table size and circuit area.

7.5 Single-Base Domain

To solve the problem of efficient table addressing in the classical method, we derive a direct mapping between the representations of a single-digit MDLNS and a redundant LNS (where different exponents imply the same value). This is done by generating a new exponent, v, or single-base domain (SBD), as shown in Equation (7.12), which is formed from the single-digit MDLNS exponents. We choose the same base from the MDLNS representation with the unconstrained exponent range, D_1 (2 in this case), as it will minimize the table sizes.

$$D_1^v = \prod_{j=1}^{k} D_j^{a_j} \tag{7.12}$$

Equation (7.12) can be manipulated through logarithms to obtain Equation (7.13).

$$v = \sum_{j=1}^{k} a_j \cdot \log_{D_1}(D_j) \tag{7.13}$$

Since v is most often a real number, for hardware implementation it should be represented in an integer form by using a fixed-point representation where a limited number of bits is used to represent the fractional part of the real number (i.e., $v = v_i + (v_f/m)$, $m = 2^r$). The integer portions of v are shifted by r bits and are combined with the fractional part so that arithmetic functions operate with an effective decimal point; v_i and v_f in effect become a single word. In application-specific fixed-point systems, the number of bits for the fractional representation is generally optimized for the particular application. This is done to minimize hardware but at a cost of some quantization loss in the representation, especially if m is small. In the case of the SBD, m may not necessarily need to be very large to accommodate the factional representation since the SBD is as redundant as the MDLNS. This offers flexibility in the generation of addition and subtraction tables as well as in their reductions, which we will see later. It is important to note that the SBD is not intended for the simple operations such as multiplication and division, as it will generate improper results due to its inherent redundancy. Statistically, we have found that the number of bits required for the SBD exponent (when m is a power of 2) is 4.46, 6.36, and 7.18 more than that of the MDLNS exponents for $k = \{2, 3, 4\}$, respectively. The physical representation is also a special two-word system, with the integer portion implemented in a two's complement form, while the fraction is unsigned, but is understood to have the same sign as the integer part.

7.5.1 Single-Digit MDLNS to Single-Base Domain

The value of m is selected in order to generate accurate addition and subtraction tables, which will be shown later. We require an efficient hardware implementation to convert a set of single-digit MDLNS exponents into a single-base exponent. To keep the circuit simple, we refer to the SBD value as being $v \cdot m$.

$$v \cdot m = v_i \cdot m + v_f \tag{7.14}$$

Equations (7.13) and (7.14) can be used to determine the SBD exponent, v, and subsequently the SBD value, $v \cdot m$, from the core MDLNS sequence (the sequence of MDLNS number representations between 1 and 2). Table 7.2 shows the generated SBD exponents for $k = 3$, $D_j = 2, 3, 5$, $a_j = \{-16, \ldots, 15\}, \{-1, 0, 1\}, \{-1, 0, 1\}$, and $m = 128$. Additional entries (shaded) are shown to demonstrate the cyclical properties of the MDLNS representation. These specifications will be used in subsequent examples throughout this chapter. In order to show full tables in detail, we have limited the range of $\{b_2, b_3, \ldots, b_k\}$; however, this will result in a very coarse representation space. The table sizes presented in the following examples will strongly depend on the choice of bases as well as the exponent ranges. We therefore cannot expect the results of these examples to be typical of the MDLNS in general.

Table 7.2 demonstrates that there is a corresponding unique v_f for every combination of $a_2, a_3, \ldots, a_k$. As this sequence repeats, generating a larger MDLNS representation, a_1 and v_i increase by 1. Similarly, a_1 and v_i decrease by 1 as the sequence repeats, generating smaller MDLNS representations. Therefore the v_i component for any set of MDLNS exponents is the difference between the first exponent, b_1, and the associate core MDLNS first exponent, a_1. Using this relationship, Table 7.2 can be reordered and referenced by $a_2, a_3, \ldots, a_k$ (in binary) to find v_f and to compute v_i for any set of single-digit MDLNS exponents (see Table 7.3). This process can be easily implemented in

TABLE 7.2

SBD Exponents of the Core MDLNS Sequence for $k = 3$, $D_j = 2, 3, 5$, $a_j = \{-16, \ldots, 15\}, \{-1, 0, 1\}, \{-1, 0, 1\}$ and $m = 128$

a_1	a_2	a_3	$\prod_{j=1}^{k} D_j^{a_j}$	v_i	v_f
−1	1	1	0.937500	−1	116
0	0	0	1.000000	0	0
4	−1	−1	1.066667	0	12
1	1	−1	1.200000	0	34
−2	0	1	1.250000	0	41
2	−1	0	1.333333	0	53
−1	1	0	1.500000	0	75
3	0	−1	1.600000	0	87
0	−1	1	1.666667	0	94
−3	1	1	1.875000	0	116
1	0	0	2.000000	1	0

TABLE 7.3

MDLNS-to-SBD Conversion LUT for $k = 3$, $D_j = 2, 3, 5$, $a_j = \{-16, \ldots, 15\}, \{-1, 0, 1\}, \{-1, 0, 1\}$, and $m = 128$

Input		Output	
a_2	a_3	a_1	v_f
0	0	0	0
0	1	–2	41
0	–1	3	87
1	0	–1	75
1	1	–3	116
1	–1	1	34
–1	0	2	53
–1	1	0	94
–1	–1	4	12

hardware with a LUT. Note that the complexity of this LUT is based on R in Equation (7.3) since it does not depend in any way on the range of a_1.

Example 7.1

Given $b_j = (5, 1, -1)$, find $v \cdot m$ using both Equation (7.13) and Table 7.3.

Solution (Equation Method)

- From Equation (7.13): $v \cdot m = (a_1 + a_2 \times \log_2(3) + a_3 \times \log_2(5)) \times m = 545.67$.

Solution (Table Method)

- Looking up $a_{2\to k} = (1, -1)$ in Table 7.3 returns $v_f = 34$ and $b_1 = 1$.
- $v_i = b_1 - a_1 = 5 - 1 = 4$.
- $v \cdot m = v_i \cdot m + v_f = 4 \cdot 128 + 34 = 546$.

7.5.2 Single-Base Domain to Single-Digit MDLNS

The process of converting a single-base exponent to a set of single-digit MDLNS exponents can be achieved using the same technique as in converting a binary value into single-digit 2DLNS [9], that is, through table reversal. The originating process uses a LUT referenced by $a_2, a_3, \ldots, a_k$ to obtain a_1 (v_i offset) and v_f. We reverse the LUT so that it will be addressed only by v_f to obtain a_j. As the range of v_f is larger than that of $a_2, a_3, \ldots, a_k$, the new LUT will contain undefined entries along with a new entry to maintain the cyclical connectivity of the core MDLNS sequence (see Table 7.4).

The undefined entries are filled with the nearest single-digit MDLNS representation corresponding to v (see Table 7.5). This process introduces

TABLE 7.4

Preliminary SBD-to-MDLNS Conversion LUT for $k = 3$, $D_j = 2, 3, 5$, $a_j = \{-16, \ldots, 15\}$, $\{-1, 0, 1\}$, $\{-1, 0, 1\}$, and $m = 128$

Input	Output		
v_f	a_1	a_2	a_3
0	0	0	0
12	4	−1	−1
34	1	1	−1
41	−2	0	1
53	2	−1	0
75	−1	1	0
87	3	0	−1
94	0	−1	1
116	−3	1	1
128	1	0	0

TABLE 7.5

Complete SBD-to-MDLNS Conversion LUT for $k = 3$, $D_j = 2, 3, 5$, $a_j = \{-16, \ldots, 15\}$, $\{-1, 0, 1\}$, $\{-1, 0, 1\}$, and $m = 128$

Input	Output		
v_f	a_1	a_2	a_3
$0 \to 6$	0	0	0
$7 \to 23$	4	−1	−1
$24 \to 37$	1	1	−1
$38 \to 47$	−2	0	1
$48 \to 64$	2	−1	0
$65 \to 80$	−1	1	0
$81 \to 90$	3	0	−1
$91 \to 105$	0	−1	1
$106 \to 122$	−3	1	1
$123 \to 127$	1	0	0

redundancy into the SBD and also eliminates the last entry originally added for cyclical connectivity.

In general, depending on the value of m, it may be inefficient to implement this table with a LUT, as there will only be $R + 1$ unique outputs but 2^m entries. Most memory generators create memories with only power of 2 rows, resulting in potentially large LUTs with a considerable number of duplicate outputs. A RALUT (introduced in Chapter 6) implements these data more efficiently, as they only contain one entry per unique row, in this case $R + 1$.

Example 7.2

Given $v \cdot m = 343$ ($v_i = 2$ and $v_f = 87$), find b_j using Table 7.5 and verify with Equation (7.13).

Solution

- Looking up $v_f = 87$ in Table 7.5 returns $a_j = (3, 0, -1)$.
- $b_j = (v_i + a_1, a_2, a_3) = (5, 0, -1)$.

Verification

- From Equation (7.13): $v \cdot m = (b_1 + b_2 \times \log_2(3) + b_3 \times \log_2(5)) \times m = 342.79$.

7.5.3 MDLNS Magnitude Comparison

Since the exponent on the SBD directly relates to the magnitude of a single-digit MDLNS representation, the SBD value can be used in magnitude comparisons of MDLNS representations. For single-digit MDLNS, the SBD value of two or more representations can be easily compared to find their relative magnitudes. However, for multidigit MDLNS, there is no simple method to compute the magnitude, as there can be many combinations for a multidigit representation. In order to perform an accurate comparison, each digit of a multidigit MDLNS representation must be summed together and then have its SBD value compared.

7.6 Addition in the Single-Base Domain

Now that a monotonic relationship has been found between the values of an MDLNS representation and its exponent, we can obtain a solution for the addition operation by substituting the two SBD values for x and y into Equation (7.6). As mentioned before, the MDLNS (and the LNS) represents numbers nearest to 1 quite efficiently; approximations to $1 + D^{y-x}$ are more accurate when $x \geq y$, as this limits $\log_D(1 + D^{y-x}) \leq 1$, forcing the actual approximation to be between 1 and 2. To determine whether $x \geq y$ is trivial only requires the SBD values of x and y to be compared to each other. By always selecting the larger addend (in magnitude only) for x, the value of $x - y$ can be used to address a table providing all the solutions for Equation (7.8).

7.6.1 Computing the Addition Table

Generating the addition table by simply finding the nearest SBD approximation to satisfy Equation (7.8) will most likely generate factors that, when multiplied by the largest addend, will cause an MDLNS overflow. We can take advantage of the redundancy of the SBD to avoid the additional step for overflow correction. In fact, we can generate tables that will guarantee the best MDLNS additional result for all addend cases with no chance of overflow. The approach is simple: compute the multiplying factor from the sum of all the possible MDLNS combinations of addends for a particular value of m. If, during this process, an inaccuracy occurs such that two different MDLNS factors occupy the same entry in the table $(x - y)$, the process is restarted for a new value of m. Since this process is being computed in the SBD, there is a considerable degree of redundancy in the representation such that an addition LUT can easily be generated. For example, from Table 7.5 we can see that a wide range of SBD values can be equivalent to a single MDLNS representation; results between 7 and 23, in this case, are equivalent since they yield the same MDLNS representation. This provides us with a wide range of acceptable solutions that deliver accurate results. Furthermore, the redundancy of the SBD can be further exploited since the two inputs, x and y, can generate several possible values of $x - y$, which allows for greater deviation in the end result. In terms of Equation (7.8), we can have several table entries for $x - y$ and several solutions for each.

7.6.1.1 Minimizing the Search Space

Generating factors based on all the possible addends can be a consuming process; however, not all the combinations of addends need to be considered. Since the MDLNS and SBD share a common base 2, any multiple of 2 applied to the addends will also apply to the sum. Therefore, all of these duplicates from the total search space can be omitted (e.g., $1 + 2 = 3$ and $2 + 4 = 6$ are duplicates). The cases where one addend is considerably larger than the other can also be omitted since the multiplying factor is 1 (known as essential zero [10]). The point where the factor becomes 1 can be coarsely approximated from the unity approximants of the MDLNS sequence (see Table 7.1). From this table, the nearest multiplying factor to 1.0 is 1.041667, u_1, for this single-digit MDLNS. Due to rounding, the midpoint between these two values, 1.020834, should be chosen as the smallest factor. Any factor below $(u_1 + 1)/2$ will essentially become 1. We can find the SBD exponent difference, M_A, through some manipulation:

$$\begin{aligned} 1 + D^{y-x} &< (u_1 + 1)/2 \\ D^{y-x} &< (u_1 - 1)/2 \\ D^{x-y} &> 2/(u_1 - 1) \\ x - y &> \log_D(2/(u_1 - 1)) \end{aligned} \tag{7.15}$$

For the MDLNS example earlier, if the difference between the two SBD exponents exceeds $\log_2(2/(1.041667 - 1)) = 5.584951$, or the SBD values exceeds $5.584951 \cdot m = 715$, then addition is not required since the larger addend is dominant. In practice, this value can be larger (as we will see), but we only use it to approximate the table sizes. By removing duplicate and dominant addends from the table generator, the number of calculations can be reduced considerably so that the table can be generated offline. We formalize the process of generating the addition table in Algorithm 7.1.

Algorithm 7.1

- Generate the core MDLNS sequence list, p, by finding all the possibilities given the range $b_2, b_3, \ldots, b_k$, with $b_1 = 0$ (R representations).
- Find b_1 for all the representations in p, as it is the exponent required to place the corresponding MDLNS value between 1 and 2 (normalize operation).
- Sort list p by its normalized representations.
- Set up reference for p_i, where $i > R$: $p_{i,\, i>R} = p_{i\bmod R} \times 2^{\mathrm{int}(i/R)}$.
- Step m from R upward (in power of 2 increments).
  - Create a new list $\{g\}$ with m elements for SBD mapping.
  - Step i from 0 to $m - 1$ and find the closest MDLNS representation in p to $2^{i/m}$ and save it into g_i; create mapping from p with Equation (7.13).
  - Set up function SBD (z), which maps from the MDLNS sequence to SBD using list $\{g\}$.
  - Set up function MDLNS (z), which finds the nearest MDLNS sequence of z.
  - Find M_A from Equation (7.15), increase by 10%, and round up to the next integer to avoid exceeding array boundaries.
  - Create a new list T with $M_A \times m$ entries, and initialize as empty.
  - Step i from 0 to $R - 1$.
    - $y = \mathrm{SBD}(i)$.
    - Step j from 0 to $M_A \times (R - 1)$.
      - $k = i + j$.
      - $x = \mathrm{SBD}(k)$.
      - $\Delta = x - y$.
      - If $\Delta > M_A$, skip to next loop j.
      - $f = p_k + p_i$.
      - $h_l = \mathrm{SBD}_{lowest}(\mathrm{MDLNS}(f)) - x$.
      - $h_h = \mathrm{SBD}_{highest}(\mathrm{MDLNS}(f)) - x$.
      - If T_Δ is empty,
        - $T_{\Delta,l} = h_l$
        - $T_{\Delta,h} = h_h$
        - Skip to next loop j.
      - If $h_l > T_{\Delta,l}$, $T_{\Delta,l} = h_l$.

        - If $h_h < T_{\Delta,h}$, $T_{\Delta,h} = h_h$.
        - If $T_{\Delta,l} < 0$, $T_{\Delta,l} = 0$.
        - If $T_{\Delta,l} > T_{\Delta,h}$, abort table generation, and skip to next loop *m*.
    - End loop *j*.
        - End loop *i*.
    - Reduce table *T* and save dimensions.
- End loop *m*.
- Use smallest reduced table for implementation.

7.6.1.2 Table Redundancy

When each entry in the table is generated, the corresponding factor is actually a range of possible factors since the SBD introduces redundancy in the addition calculation. From Table 7.5, only 10 of the 128 possible SBD values are unique. Table 7.6 shows (in three columns) the complete addition table for $m = 128$; both values have been multiplied by *m* to generate only integer values. No other power of 2 lower than 128 could be chosen to generate the best representation table. This table ranges from 0 to 889 but only contains 175 elements. To minimize the table output, the lowest of the factor range will be used to generate a 889 × 7 LUT. The number of output bits is the number of bits needed to represent *m*. To keep the LUT as small as possible, values above 889 will return 0 using special logic.

7.6.1.3 RALUT Implementation

As only 175 of the 889 entries of the table are used, this LUT could easily be directly implemented into a RALUT to reduce its size (175 × 7 RALUT). The logic required in the LUT implementation to compensate for inputs above 889 is already incorporated into the RALUT.

7.6.1.4 Table Merging

It is possible to achieve additional table reductions in the RALUT by merging adjacent table entries. From Table 7.6 entry 0 to 6 contains a common factor range of 125 to 131, entry 7 to 11 a common factor range of 115 to 134, and entry 12 to 18 a common factor range of 113 to 131. These three entries can be merged into a single entry from 0 to 18 with a range of 125 (the largest low value) to 131 (the smallest high value). The next entry, 19 to 21, has a high factor of 122, which is lower than any of the previous low factors and therefore could not have been included in this group merging. We can continue performing this process with the remaining entries to further reduce the table. When completed, the lowest factor is selected to minimize the required bits on the output. For this example, the 889 × 7 LUT or 175 × 7 RALUT can be reduced into a 23 × 7 RALUT (see Table 7.7).

TABLE 7.6

Complete Addition LUT for $k = 3$, $D_j = 2, 3, 5$, $a_j = \{-16, \ldots, 15\}, \{-1, 0, 1\}, \{-1, 0, 1\}$, and $m = 128$

$x - y$	Factor	$x - y$	Factor	$x - y$	Factor	$x - y$	Factor
0	125 → 131	227	48 → 52	449	12 → 25	674	4 → 5
7	115 → 134	232	36 → 49	452	14 → 30	681	0 → 5
12	113 → 131	234	41 → 52	459	14 → 15	686	4 → 11
19	115 → 122	237	31 → 46	466	12 → 18	693	4 → 6
22	103 → 131	244	40 → 47	471	14 → 15	700	0 → 11
24	111 → 122	249	36 → 46	478	12 → 23	703	0 → 5
29	111 → 117	256	41 → 46	483	12 → 18	705	0 → 11
34	106 → 117	263	29 → 49	488	7 → 18	708	4 → 13
41	106 → 110	268	38 → 47	490	12 → 25	715	0 → 5
46	106 → 110	275	29 → 40	493	6 → 13	722	0 → 6
53	99 → 105	278	36 → 46	500	7 → 15	727	0 → 3
60	99 → 114	280	36 → 52	505	4 → 13	734	0 → 6
63	93 → 100	285	36 → 39	512	7 → 13	739	0 → 3
65	99 → 110	290	31 → 35	519	7 → 11	744	0 → 6
68	89 → 100	297	31 → 35	524	7 → 11	746	0 → 11
75	94 → 100	302	31 → 35	531	0 → 11	749	0 → 3
82	82 → 92	309	29 → 35	534	7 → 13	756	0 → 5
87	89 → 90	316	29 → 40	536	0 → 11	761	0 → 3
94	82 → 90	319	19 → 30	541	7 → 18	768	0 → 3
99	79 → 88	321	28 → 35	546	7 → 13	775	0 → 6
104	77 → 92	324	14 → 30	553	7 → 11	780	0 → 3
106	82 → 93	331	24 → 30	558	4 → 11	787	0 → 11
109	77 → 88	338	24 → 28	565	7 → 11	790	0 → 3
116	77 → 80	343	26 → 27	572	0 → 11	792	0 → 11
121	72 → 78	350	24 → 27	575	7 → 15	797	0 → 6
128	72 → 76	355	19 → 25	577	0 → 11	802	0 → 3
135	72 → 81	360	19 → 35	580	4 → 13	809	0 → 5
140	70 → 78	362	12 → 25	587	7 → 13	814	0 → 3
147	58 → 69	365	16 → 30	594	7 → 11	821	0 → 3
150	60 → 71	372	19 → 30	599	7 → 11	828	0 → 11
152	53 → 68	377	19 → 30	606	7 → 11	831	0 → 5
157	65 → 76	384	19 → 23	611	4 → 11	833	0 → 11
162	60 → 71	391	12 → 23	616	7 → 18	836	0 → 3
169	60 → 64	396	19 → 23	618	0 → 11	843	0 → 3
174	57 → 64	403	12 → 27	621	6 → 13	850	0 → 6
181	58 → 64	406	19 → 30	628	0 → 15	855	0 → 3
188	41 → 71	408	12 → 25	633	4 → 13	862	0 → 6
191	50 → 59	413	19 → 23	640	4 → 5	867	0 → 3
193	53 → 68	418	16 → 18	647	0 → 6	872	0 → 6
196	47 → 56	425	16 → 18	652	4 → 6	874	0 → 11

TABLE 7.6 (CONTINUED)

Complete Addition LUT for $k = 3$, $D_j = 2, 3, 5$, $a_j = \{-16, \ldots, 15\}, \{-1, 0, 1\}, \{-1, 0, 1\}$, and $m = 128$

$x - y$	Factor	$x - y$	Factor	$x - y$	Factor	$x - y$	Factor
203	50 → 56	430	14 → 18	659	0 → 11	877	0 → 3
210	50 → 57	437	12 → 18	662	4 → 5	884	0 → 5
215	48 → 52	444	12 → 28	664	0 → 11	889	0 → 3
222	48 → 49	447	7 → 18	669	0 → 6		

TABLE 7.7

Complete Addition RALUT for $k = 3$, $D_j = 2, 3, 5$, $a_j = \{-16, \ldots, 15\}$, $\{-1, 0, 1\}, \{-1, 0, 1\}$, and $m = 128$

$x - y$	Factor	$x - y$	Factor	$x - y$	Factor
0 → 18	125	147 → 168	65	343 → 354	26
19 → 40	115	169 → 190	60	355 → 417	19
41 → 52	106	191 → 214	53	418 → 458	16
53 → 81	99	215 → 236	48	459 → 492	14
82 → 98	89	237 → 274	41	493 → 639	7
99 → 115	82	275 → 289	36	640 → 726	4
116 → 127	77	290 → 318	31	727 → 1,023	0
128 → 146	72	319 → 342	28	—	—

Example 7.3

In MDLNS, find the sum of $x_j = (5, 1, -1) = 19.2$ and $y_j = (5, 0, -1) = 6.4$.

Solution

- Find the SBD exponent of the first addend from Table 7.3: $v_{i_x} = 4$ and $v_{f_x} = 34$.
- Find the SBD exponent of the second addend: $v_{i_y} = 2$ and $v_{f_y} = 87$.
- First addend is larger, no reordering necessary.
- Find SBD difference: $v_f = v_{f_x} - v_{f_y} = 34 - 87 = 75$ (modulo 128) and $borrow = 1$, $v_i = v_{i_x} - v_{i_y} - borrow = 4 - 2 - 1 = 1$.
- Look up $v_i \times m + v_f = 203$ in Table 7.7: factor = 53 ($v_{i_a} = 0$ and $v_{f_a} = 53$).
- Add largest addend to factor: $v_{f_s} = v_{f_x} + v_{f_a} = 34 + 53 = 87$ (modulo 128) and $carry = 0$, $v_{i_s} = v_{i_x} + v_{i_a} + carry = 4 + 0 + 0 = 4$.
- Find the MDLNS representation of the sum from Table 7.5: $b_j = (3, 0, -1)$.
- $z_j = (v_{i_s} + b_1, b_2, b_3) = (7, 0, -1)$.
- The MDLNS sum is $2^7 \times 3^0 \cdot 5^{-1} = 25.6$.
- The actual sum is $19.2 + 6.4 = 25.6$ ($\Delta = 0$).

Verification

- The two nearest approximations to $2^7 \times 3^0 \times 5^{-1}$ are $2^3 \times 3^1 \times 5^0 = 24$ ($\Delta = 1.6$) and $2^4 \times 3^{-1} \times 5^1 = 26.67$ ($\Delta = 1.067$).
- $2^7 \times 3^0 \times 5^{-1} = 25.6$ is the best solution for the given system.

7.6.1.5 Alternatives for m

As the addition tables are being generated, some addend combinations may produce factors that are outside the existing ranges, resulting in an inaccurate table. We subsequently change m and try the process again. Up to this point we have selected a value of m where $m = 2^r$, a rather demanding restriction. However, it is possible to obtain complete addition tables with smaller values of m. This is the case for the example shown throughout this chapter as $R = 9$, which is much smaller than the m needed to create an accurate table (i.e., 128). Values of 16, 32, and 64 did not produce such a table, but 51 values between 53 and 127 produced accurate tables with relatively the same RALUT geometries. In terms of hardware these non-power of 2 alternatives for m will require the use of modular arithmetic in the binary addition and subtraction operations, which will increase the requirements of the circuit. However, the reduction in the representation of the SBD value in the LUTs and RALUTs may offset this increase. It is this potential savings that will dictate the choice of m, either a power of 2, m_{powerof2}, or a non-power of 2, m_{any}. Both of these values are determined by the bases and the range of the associated exponents. We are unable to determine a quantitative relationship for this; however, we can compare the difference in bits using every base combination (from the first 25 prime numbers, excluding 2) for two- and three-base configurations with various exponent ranges (as shown in Figure 7.3). We

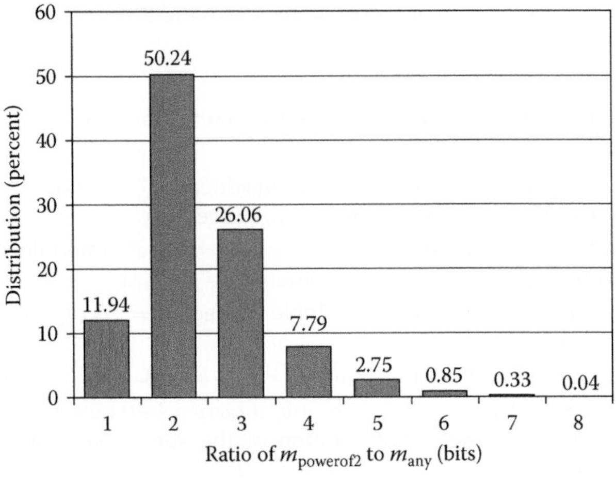

FIGURE 7.3
Histogram of the ratio between m_{powerof2} and m_{any} in bits.

can see that on average we can save 2 or more bits on the SBD value by using m_{any}. It is important to note that 3.1% of these combinations were excluded from this comparison since m_{powerof2} could not be found for less than 24 bits (due to computational resources).

7.6.2 The Complete Structure

The complete architecture for a single-digit MDLNS addition is shown in Figure 7.4. It can be easily implemented in a pipelined structure where some of the stages can be merged to reduce latency. The signs of the addends are ignored in this analysis since they have no effect on the addition operation. We will include them when we address MDLNS subtraction.

7.6.3 Results

Since MDLNS is a very generalized number system it is difficult to arrive at a derivation that relates the entries in an addition RALUT to its bases and range of exponents. We can offer a worst-case statistical estimation, Equation (7.16), based on using every base combination (from the first 25 prime numbers, excluding 2) for $k = \{2, 3, 4\}$ with various exponent ranges.

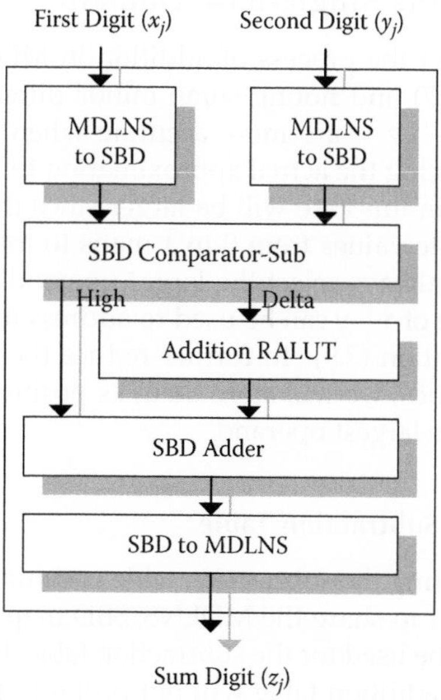

FIGURE 7.4
Single-digit MDLNS adder structure.

$$\text{ROWS}_{\text{addition,worst}} = (k{\cdot}2.89 - 2.28){\cdot}R \tag{7.16}$$

Based on these sample combinations, the rows in the RALUT should never exceed the value from Equation (7.16). A more conservative approximation is

$$\text{ROWS}_{\text{addition,typical}} = (k{\cdot}1.55 - 0.73){\cdot}R \tag{7.17}$$

Although a formal expression has not yet been found, we have some evidence that there is a relationship between the table size and the deviation between the near-unity approximations of the current MDLNS. As the number of bases increase, the number of near-unity approximants also increase, which leads to irregular deviations in the spacing between representations in the core MDLNS sequence. We intend to further investigate this, as it may lead to more ways to exploit this number system.

A comparison between the proposed method for addition and the earlier methods is left to Section 7.8 when subtraction is also considered.

7.7 Subtraction in the Single-Base Domain

We can similarly adapt the process of addition in MDLNS for subtraction by using Equation (7.7) and noting some minor differences. The MDLNS approximations to $1 - D^{y-x}$ are more accurate when $x \geq y$, as this limits $\log_D(1 - D^{y-x}) \leq 0$, forcing the actual approximation to be between −1 and 0. The tables generated in this case will be larger since the MDLNS representation space of absolute values from 0 to 1 maps to that of 1 to the highest boundary. Again, we always select the larger operand (in magnitude only) for x, so that the value of $x - y$ can be used to address the table providing all the solutions for Equation (7.9). To further reduce the table space required to represent these values, we will store them as positive numbers and later subtract them from the largest operand.

7.7.1 Computing the Subtraction Table

The process of generating the subtraction table is similar to that of the addition table. In the effort to share the MDLNS/SBD maps, the same m for the addition table should be used for the subtraction table. It is very possible that the m selected for an addition table will not properly generate the subtraction table since its contents require slightly greater accuracy. It is generally recommended to first generate the subtraction table, and use the successful value of m to generate the addition table.

7.7.1.1 Minimizing the Search Space

Again, we can take advantage of removing duplicate and dominant operands from the table generator so that the table can be generated offline in a finite time. For subtraction, the point where the factor becomes 1 can be approximated in a similar fashion from the near-unity approximants of the MDLNS sequence. Since the factors for subtraction are between 0 and 1, we must approach this solution from below 1. From Table 7.1, the nearest multiplying factor less than 1.0 is 0.96, u_1^{-1} (or 1/1.041667) for this single-digit MDLNS. Again due to rounding, the midpoint between these two values, 0.98, should be chosen as the smallest factor. We can find the difference, M_s, in the SBD exponents through similar manipulation:

$$\begin{aligned} 1-D^{y-x} &> \left(u_1^{-1}+1\right)/2 \\ -D^{y-x} &> \left(u_1^{-1}-1\right)/2 \\ D^{y-x} &< \left(1-u_1^{-1}\right)/2 \\ D^{x-y} &> 2/\left(1-u_1^{-1}\right) \\ x-y &> \log_D\left(2/\left(1-u_1^{-1}\right)\right) \end{aligned} \tag{7.18}$$

This solution of $\log_2(2/(1-0.96)) = 5.643856$ is slightly larger than the case for addition (5.584951).

We formalize the process of generating the subtraction table in Algorithm 7.2; note that it is quite similar to Algorithm 7.1, but with some differences (bold) to avoid negative results.

Algorithm 7.2

- Generate the core MDLNS sequence list, p, by finding all the possibilities given the range $b_2, b_3, \ldots, b_k$ with $b_1 = 0$ (R representations).
- Find b_1 for all the representations in p, as it is the exponent required to place the corresponding MDLNS value between 1 and 2 (normalize operation).
- Sort list p by its normalized representations.
- Set up reference for p_i, where $i > R$: $p_{i,i>R} = p_{i\bmod R} \times 2^{\mathrm{int}(i/R)}$.
- Step m from R upward (in power of 2 increments).
    - Create a new list a with m elements for SBD mapping.
    - Step i from 0 to $m-1$ and find the closest MDLNS representation in p to $2^{i/m}$ and save it into a_i; create mapping from p with Equation (7.13).
    - Set up function SBD(z), which maps from MDLNS sequence to SBD.

- Set up function MDLNS(z), which finds the nearest MDLNS sequence of z.
- **Find M_s from Equation (7.18), increase by 10%, and round up to the highest integer to avoid exceeding array boundaries.**
- **Create a new list T with $M_s \times$ m entries, and initialize as empty.**
- **Step i from $M_s \cdot R$ to $M_s \cdot R + R - 1$.**
  - $y = \text{SBD}(i)$.
  - Step j from 0 to $M_s \cdot (R - 1)$.
    - $k = i + j$.
    - $x = \text{SBD}(k)$.
    - $\Delta = x - y$.
    - **If $\Delta > M_s$, skip to next loop j.**
    - $f = p_k - p_i$.
    - $h_h = x - \text{SBD}_{lowest}$ **(MDLNS(f)).**
    - $h_l = x - \text{SBD}_{highest}$ **(MDLNS(f)).**
    - If T_Δ is empty,
      - $T_{\Delta, l} = h_l$
      - $T_{\Delta, h} = h_h$
      - Skip to next loop j.
    - If $h_l > T_{\Delta, l}$, $T_{\Delta, l} = h_l$.
    - If $h_h > T_{\Delta, h}$, $T_{\Delta, h} = h_h$.
    - If $T_{\Delta, l} < 0$, $T_{\Delta, l} = 0$.
    - If $T_{\Delta, l} > T_{\Delta, h}$, abort table generation and skip to next loop m.
  - End loop j.
- End loop i.
- Reduce table T and save dimensions.

- End loop m.
- Use smallest reduced table for implementation.

7.7.1.2 RALUT Implementation and Table Reduction

Using the example specifications, the resulting subtraction table (not shown) can be implemented in a 889 × 10 LUT. This table has the same number of rows as that of the addition table, as they use the same operand combinations during the table generation algorithms. The output word is larger since the integer portion of the SBD exponent is greater than 1. The number of output bits is the sum of the number of bits needed to represent M_s and m. The subtraction table contains a special first entry to minimize the logic to compensate for the condition when the SBD value difference is zero. This same LUT can be implemented in a RALUT to achieve the same rows as that of the addition table (175 × 10). To further minimize the table size, the row merging procedure can be applied to obtain a 41 × 10 RALUT (see Table 7.8) that contains the same information as the other two tables. This RALUT is larger than that of the addition operation since the MDLNS represents values between 0 and 1 more accurately (i.e., more representations) than values between 1 and 2.

TABLE 7.8

Complete Subtraction RALUT for $k = 3$, $D_j = 2, 3, 5$, $a_j = \{-16, \ldots, 15\}$, $\{-1, 0, 1\}$, $\{-1, 0, 1\}$, and $m = 128$

$x - y$	Factor	$x - y$	Factor	$x - y$	Factor
0 → 6	1023	87 → 93	180	278 → 308	40
7 → 11	591	94 → 103	166	309 → 318	38
12 → 18	509	104 → 115	154	319 → 330	36
19 → 21	414	116 → 120	139	331 → 361	30
22 → 28	395	121 → 127	135	362 → 383	26
29 → 33	349	128 → 134	125	384 → 417	23
34 → 40	328	135 → 146	117	418 → 448	18
41 → 45	294	147 → 168	103	449 → 470	16
46 → 52	279	169 → 173	93	471 → 511	14
53 → 59	253	174 → 195	82	512 → 540	11
60 → 67	228	196 → 214	72	541 → 651	7
68 → 74	210	215 → 231	69	652 → 733	4
75 → 81	200	232 → 248	60	734 →1,023	0
82 → 86	185	249 → 277	52	—	—

Example 7.4

In MDLNS, find the difference between $y_j = (5, 0, -1) = 6.4$ and $x_j = (5, 1, -1) = 19.2$.

Solution

- Find the SBD exponent of the minuend from Table 7.3: $v_{i_y} = 2$ and $v_{f_y} = 87$.
- Find the SBD exponent of the subtrahend: $v_{i_x} = 4$ and $v_{f_x} = 34$.
- The minuend is smaller; swap minuend and subtrahend, indicate a sign change.
- Find SBD difference: $v_f = v_{f_x} - v_{f_y} = 34 - 87 = 75$ (modulo 128) and $borrow = 1$, $v_i = v_{i_x} - v_{i_y} - borrow = 4 - 2 - 1 = 1$.
- Look up $v_i \times m + v_f = 203$ in Table 7.8: factor = 72 ($v_{i_s} = 0$ and $v_{f_s} = 72$).
- Subtract factor from largest input: $v_{f_s} = v_{f_x} - v_{f_s} = 34 - 72 = 90$ (modulo 128) and $borrow = 1$, $v_{i_s} = v_{i_x} - v_{i_s} - borrow = 4 - 0 - 1 = 3$.
- Find the MDLNS representation of the difference from Table 7.5: $b_j = (3, 0, -1)$.
- $w_j = (v_{i_s} + b_1, b_2, b_3) = (6, 0, -1)$.
- The MDLNS difference is $-1 \times 2^6 \times 3^0 \times 5^{-1} = -12.8$.
- The actual difference is $6.4 - 19.2 = -12.8$ ($\Delta = 0$).

Verification

- The two nearest approximations to $-1 \times 2^6 \times 3^0 \times 5^{-1}$ are $-1 \times 2^2 \times 3^1 \times 5^0 = -12$ ($\Delta = 0.8$) and $-1 \times 2^3 \times 3^{-1} \times 5^1 = 13.33$ ($\Delta = 0.53$).
- $-1 \times 2^6 \times 3^0 \times 5^{-1} = -12.8$ is the best solution for the given system.

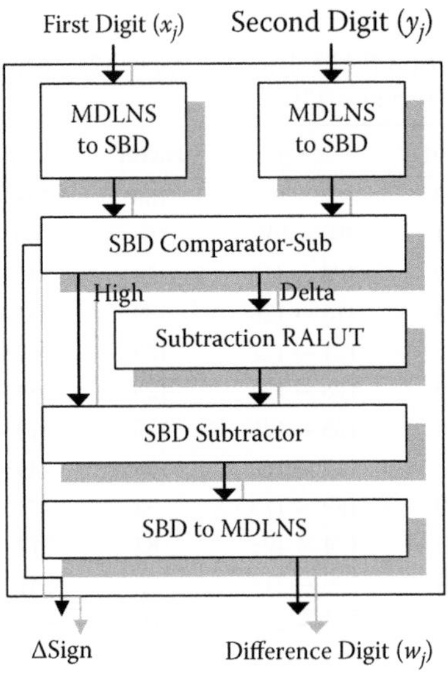

FIGURE 7.5
Single-digit MDLNS subtractor structure.

7.7.2 The Complete Structure

The complete structure for a single-digit MDLNS subtraction is shown in Figure 7.5. It can also be implemented in a pipelined structure similar to the addition process. Again, the signs of the input are ignored for now; however, an additional output is used to indicate if the result is negated. The sign will be handled in the next section.

7.7.3 Results

Similarly to the addition method, we can estimate a worst and typical case relationship between the RALUT entries and R based on statistical observations. In general there appears to be a relatively constant ratio of 1.5 between the number of entries in the subtraction to addition RALUTs.

$$\text{ROWS}_{\text{subtraction,worst}} = (k \cdot 4.34 - 3.42) \cdot R \tag{7.19}$$

$$\text{ROWS}_{\text{subtraction,typical}} = (k \cdot 2.36 - 1.1) \cdot R \tag{7.20}$$

We shall use these expressions, along with those from the addition tables, to compare the proposed method with prior methods in the next section.

7.8 Single-Digit MDLNS Addition/Subtraction

If the value of m is the same in both the addition and subtraction tables, the circuits can be merged to create a full adder/subtractor circuit by sharing common components such as the MDLNS/SBD maps, SBD addition/subtraction, etc.

7.8.1 The Complete Structure

With the addition of a multiplexer and some logic to determine the operation and output sign, the full circuit can be realized (see Figure 7.6).

7.8.2 Results

At this point, we have generated two worst-case approximations for the rows in the addition and subtraction RALUTs for $k = \{2, 3, 4\}$. We will also assume the worst-case value for R such that $R = 2^{C_k}$, although in practice it can be smaller. Table 7.9 shows a comparison between the number of rows and output bits needed for the pure LUT, classical, and proposed methods; it also includes the dimensions of the MDLNS/SBD maps. The pure LUT and classic tables have been separated into two tables to simplify the comparison.

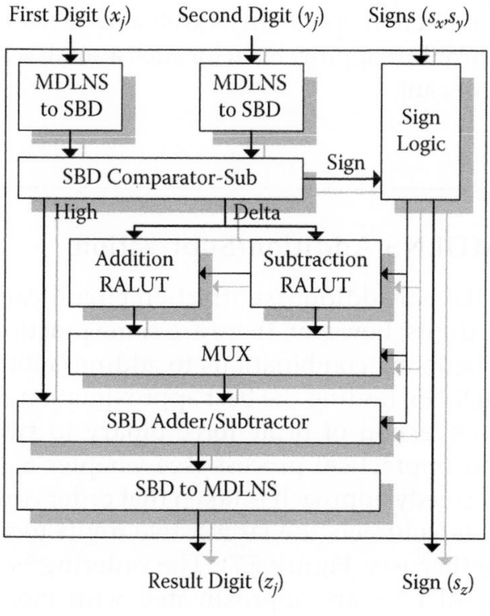

FIGURE 7.6
Single-digit MDLNS adder/subtractor structure.

TABLE 7.9

LUT and RALUT Requirements for Existing and Proposed (Worst-Case) Method

Method	Rows	Output Bits
Pure LUT addition	$2^{2(C_1+C_{2\to k}+1)}$	$C_1 + C_{2\to k} + 1$
Pure LUT subtraction	$2^{2(C_1+C_{2\to k}+1)}$	$C_1 + C_{2\to k} + 1$
Classic (LUT) addition	$2^{(C_1+C_{2\to k}+k)}$	$(C_1 + C_{2\to k} + k) + 1$
Classic (LUT) subtraction	$2^{(C_1+C_{2\to k}+k)}$	$(C_1 + C_{2\to k} + k) + 2$
Proposed addition (RALUT)	$2^{C_{2\to k}} \cdot (k \cdot 2.89 - 2.28)$	$\log_2(m)$
Proposed subtraction (RALUT)	$2^{C_{2\to k}} \cdot (k \cdot 4.34 - 3.42)$	$\log_2(M_s) + \log_2(m)$
MDLNS-to-SBD mapper (LUT)	$2^{C_{2\to k}}$	$C_1 + \log_2(m)$
SBD-to-MDLNS mapper (RALUT)	$C_{2\to k}$	$C_1 + C_{2\to k}$

Both the pure LUT and classical methods depend heavily on C_1 and $C_{2\to k}$ such that the table sizes quadruple and double, respectively, as either C increases by 1 bit. The proposed method only depends on $C_{2\to k}$, as this value affects the resolution of the MDLNS representation. The output bits are also reduced in the proposed method, as it only depends on m and the range for which MDLNS subtraction is effective. The MDLNS/SBD mappers add some overhead to the system, but this is acceptable considering the savings between proposed and earlier methods.

For the example system shown throughout the chapter, the pure LUT method requires a 20 Mbit LUT, while the classical method requires a 108 kbit LUT. The proposed approach requires a 23 × 7 and 41 × 10 RALUT for the addition and subtraction operation as well as a 16 × 9 LUT and a 10 × 6 RALUT for MDLNS/SBD mapping. Even for such a small scale, the hardware savings is very significant.

7.9 Two-Digit MDLNS Addition/Subtraction

The single-digit MDLNS addition/subtraction circuit can be extended to operate with more digits; however, there are some practical limitations. In general there can be many combinations to adding/subtracting four values together. In MDLNS, finding the best approximation can be compared to performing a conversion of brute-force binary to two-digit MDLNS, a very lengthy and impractical process (see Chapter 6). To simplify the implementation, a greedy approach is taken that orders (in magnitude) the four digits and adds/subtracts the largest two digits together and smallest two digits together (see Figure 7.7). The ordering is important since smaller values in MDLNS are approximated with more accuracy than larger values. Experimentally, the sum/difference operation is more accurate when adding/subtracting digits that are relatively similar in absolute magnitude compared to digits in any order. These empirical results are

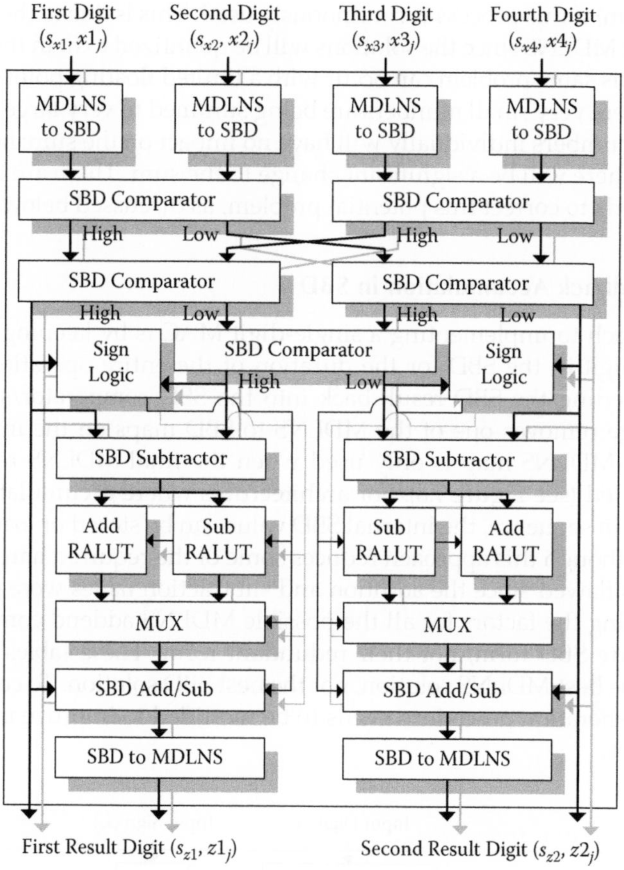

FIGURE 7.7
Two-digit MDLNS adder/subtractor structure.

in agreement with similar, well-established observations about floating point summation [11].

7.10 MDLNS Addition/Subtraction with Quantization Error Recovery

Using the single-digit MDLNS addition/subtraction process, as described above, operates well for finding the nearest MDLNS sum/difference between two or more MDLNS digits. However, in system architectures (e.g., multiply and accumulate, or MAC, in finite impulse response (FIR) filters) where a large number of addition/subtraction operations occur, there can be a considerable loss in numerical accuracy (depending on the ranges of the exponents in the system) if this process is continuously used. This is due to the nature of single-digit MDLNS since the solutions will be quantized to fit in the MDLNS domain. This same problem can occur with a limited floating point system or an LNS where very small numbers are being summed to very large numbers. The small numbers individually will have no impact on the sum; however, if combined there will be a significant change to the sum. There are three possible methods to correct this potential problem, as discussed below.

7.10.1 Feedback Accumulation in SBD

One approach to implementing a single-digit MAC is by keeping the accumulation digit in the SBD for the duration of the entire operation. This is done by feeding the SBD result back into the SBD comparator/subtractor. This change removes one of the MDLNS-to-SBD maps on the input while the SBD-to-MDLNS map is only used when the final MDLNS representation is needed (see Figure 7.8). For architectures where accumulation is not performed in sequence, the internal SBD value can be stored or loaded when needed. Although this approach reduces some of the required hardware, the accuracy is flawed since the addition and subtraction tables were generated by computing the factors for all the possible MDLNS addend combinations in their pure SBD form, not their redundant form. These tables will only generate the best MDLNS solution, not the best SBD solution. To correct this, the table generation procedure needs to be modified to improve upon these inaccuracies.

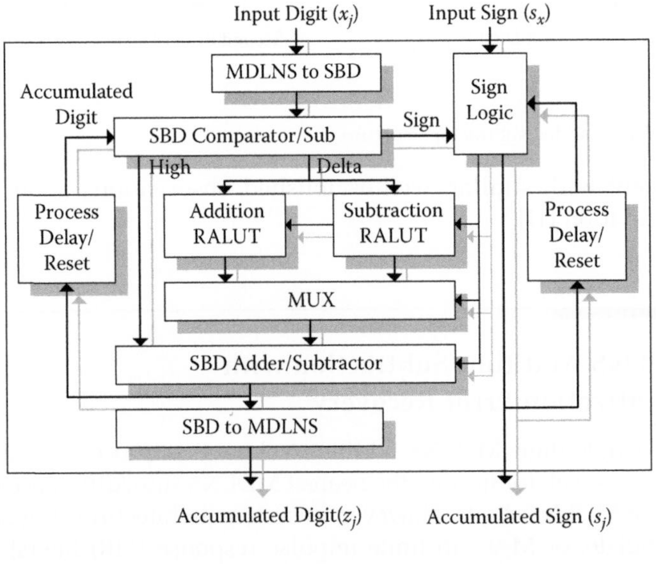

FIGURE 7.8
MDLNS MAC structure.

7.10.2 Feedback Accumulation in SBD (with Full SBD Range)

Only one of the operands (accumulation digit) needs to handle the full range (both pure and redundant) of the SBD in order to compensate for the inaccuracies present in the above method. Unfortunately this removes any redundancy for the computations with the accumulation operand and introduces many table inconsistencies that need to be corrected to generate accurate tables. We have developed a preliminary procedure to consider all the possibilities in order to correct these tables; however, this procedure is computationally intensive and it has, thus far, only created tables in specific instances where the exponent ranges are very small.

7.10.3 Increasing the SBD Accuracy Internally

A simpler and more viable solution is to increase the accuracy of the SBD in only the accumulation circuit. This concept is similar to using internal bit widths that are larger than the inputs and outputs in binary multipliers to increase numerical accuracy effectively when handling fractional multiplicands. To do this we simply select a larger range for the MDLNS exponents and generate the addition and subtraction tables with the nonaccumulation digit set to the same accuracy as the input and output. The result from either operation will most likely be in a redundant SBD form, which therefore needs to be mapped to a pure SBD form; this is done with a RALUT with R entries (see Figure 7.9). The SBD/MDLNS maps would use the original exponent

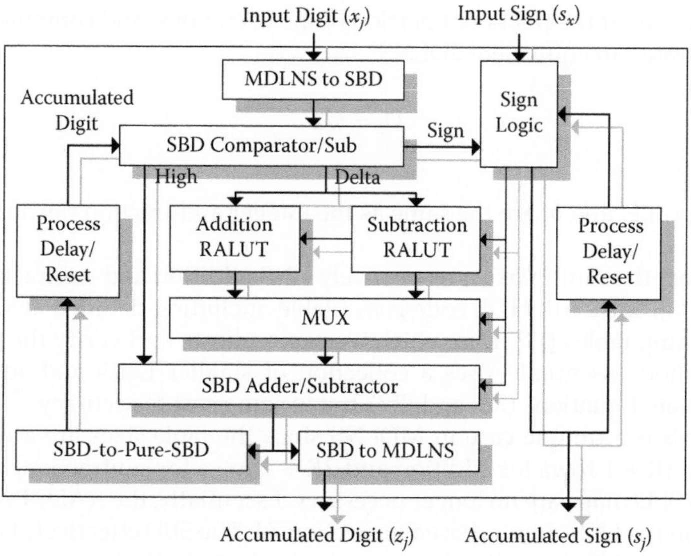

FIGURE 7.9
MDLNS MAC with accuracy correction.

ranges so there would be no additional hardware associated with them. The main advantage of this method is that the tables are relatively small and they can still be generated very quickly. In terms of the table sizes, with this approach we have statistically determined, regardless of the selection of the bases, that if the internal accumulation circuit has an accuracy set to R and an input/output accuracy of $R/2$, the table sizes will be approximately 93%, 77%, and 60% (with $k = \{2, 3, 4\}$, respectively) of the original if the input/output accuracy is also R.

7.11 Comparison to an LNS Case

The proposed method for addition and subtraction in MDLNS cannot truly be compared to any other method as there are no other methods in published literature. It is well known how to characterize the overall impact of errors when arbitrary numbers are converted to the same-sized floating point and equivalent conventional LNS representations, thus allowing fair comparisons between floating point and LNS [2,4–8,12–14]. It is more difficult to compare the proposed addition and subtraction algorithms, for an arbitrary MDLNS, with one-dimensional arithmetic systems, because the accuracy of MDLNS is more difficult to characterize due to its multidimensional characteristics. It is possible, however, to compare published LNS addition and subtraction methods because LNS is a subset of MDLNS. To do this we first select a range for the fractional portion of the LNS, r bits, and compute a base that generates an equivalent 2DLNS:

$$D_2 = \sqrt[2^r]{2} \tag{7.21}$$

The range of b_1 and b_2 are the same as the integer and fraction portion of the LNS.

We select the multipartite, respectively LNS addition and subtraction circuits in [13], since full HDL code is available, including detailed algorithms for obtaining tables [12] from which we can replicate and verify the results. This method essentially uses a collection of smaller LUTs and adders to approximate Equations (7.6) and (7.7), but at some cost to accuracy.

The LNS is a unique case in MDLNS since the table sizes are accurately predicted ($R + 1$ rows for addition and $2R + 1$ rows for subtraction) and the MDLNS/SBD maps are no longer necessary. Essentially, the RALUT replaces the LUT in the LNS circuits based on Section 7.4. The SBD effectively becomes a fixed-point LNS since all redundancy is eliminated. We therefore will only consider the complexities of the RALUTs and LUTs since these are the only differences in the circuitry between the LNS and MDLNS.

7.11.1 Addition

Table 7.10 shows a comparison between the addition tables using the multipartite method in [13] and the proposed method parameterized for an equivalent LNS. The proposed method always delivers the best solution, but at a significant cost in hardware, especially as r increases. The multipartite method is superior in requiring lower circuit area, but it cannot perfectly encode the addition tables because of the requirement that the slope of the function should change very little. The accuracy is always above 90%, and it deviates by no more than 1 unit (a factor of $\sqrt[2^r]{2}$). We can equally encode the tables prior to reduction (e.g., Table 7.6) using the multipartite method and obtain similar results. Our goal here, however, is to obtain 100% accuracy for any single-digit MDLNS. If we change the bases and the ranges on exponents, such as those used in the examples throughout this chapter, the addition function will not have a slowly increasing slope, such as that in Figure 7.1. We therefore cannot easily use the multipartite approach for encoding the tables since there will be significant error; however, this may be acceptable in situations where the designer can accept some computational errors as a trade-off with hardware savings.

7.11.2 Subtraction

Table 7.11 shows a comparison between the subtraction tables using the methods from the previous subsection. In this case, the multipartite method requires a multisystem solution with multiple tables since the absolute slope of the function increases rapidly as it approaches 0 (i.e., Figure 7.2). The accuracy is much lower than that of the addition tables, and although no more than 1% of the solutions deviate by more than 4 units, the maximum deviation is very large and can generate unexpected results for particular operands. The authors intentionally added this error to compensate for the effect of "catastrophic cancellation" [14], which affects any LNS subtraction

TABLE 7.10
Comparison of LNS Addition Tables for Various Values of r

	Multipartite Method [13]			Proposed Method	
r	No. of LUTs	Total Bits	Accuracy (%)	RALUT	Total Bits
6	2	768	91.8	65 × 7	455
7	4	1,074	95.4	129 × 8	1,032
8	4	1,908	94.0	257 × 9	2,313
9	4	2,380	90.3	513 × 10	5,130
10	4	4,636	94.7	1,025 ×11	11,275
11	5	6,433	94.1	2,049 × 12	24,588
12	5	11,164	94.0	4,097 × 13	53,261
13	5	16,613	96.4	8,193 × 14	114,702

TABLE 7.11
Comparison of LNS Subtraction Tables for Various Values of r

	Multipartite Method [13]			Proposed Method	
r	No. of Systems, Total LUTs	Total Bits	Accuracy (%), Worst Deviation (units)	RALUT	Total Bits
6	7, 14	470	62.4, 35	129 × 9	1,161
7	8, 20	878	74.7, 60	257 × 11	2,827
8	8, 21	1,606	73.2, 89	513 × 12	6,156
9	9, 25	2,988	78.6, 177	1,025 × 13	13,325
10	10, 27	4,344	76.2, 355	2,049 × 14	28,686
11	11, 33	7,089	75.5, 709	4,097 × 15	61,455
12	12, 36	12,046	78.7, 1,418	8,193 × 16	131,088
13	13, 41	20,691	77.9, 2,836	16,385 × 17	278,545

circuit that approximates the multiplication factor. Again the proposed system is larger; however, it always delivers 100% accuracy for any single-digit MDLNS, and does not suffer from accuracy problems with operands that are close in value.

7.11.3 Comparisons

It is quite clear that an LNS implementation requires significantly more hardware resources as r increases. There has been considerable research work in the area of optimizing LNS addition/subtraction circuits. The monotonic relationship between the exponent and represented value in LNS reduces the variance in the tables, therefore allowing the multipartite approach to effectively optimize the tables with little computational error. We can, however, allow similar inaccuracies in the MDLNS addition/subtraction tables, thus reducing the hardware cost.

The real advantage of using the MDLNS is the fact that we can modify the nonbinary bases ($D_2, \ldots, D_k$) to customize the representation for particular applications without changing the complexity of the system. In Chapter 8, we provide a large example where the exponent range of a two-base MDLNS is able to be reduced by over four times by optimizing the value of D_2. This optimization process requires a set of numbers to "best represent," in this case, filter coefficients. In terms of addition and subtraction, if we can effectively reduce r by at least 2, then it would appear that the proposed system would become more competitive in size compared to the existing LNS solution, though we do not have a formal proof for such a claim. Because the table sizes can differ considerably for the selection of the bases and their associated exponent ranges, this claim would have to be tested on an individual basis for particular applications. A link to software tools that can be used for such tests is provided at the end of this chapter.

7.12 Summary

This chapter has described a hardware-realizable method of performing single-digit MDLNS addition and subtraction that generates the same solutions as with a pure LUT method for any number and choice of bases (including real bases), as well as exponent ranges. Additional digits can also be handled with the use of a magnitude sorter. Unlike the published methods, the proposed method does not depend on the exponent range of the binary base and therefore scales much better. The use of the RALUT offers a general solution for any MDLNS; however, its hardware implementation can be costly. Selecting a set of optimal bases can reduce the associated exponent ranges, thereby reducing the table sizes. Although not discussed here, it is possible to partition the table data into smaller multiple LUTs and RALUTs to attempt to reduce the overall circuit area.

Software has been written that automatically generates tables for the parameterized hardware description language (HDL) code. This code implements single-digit MDLNS addition/subtraction components for any choice of bases and range of exponents. The LUT and RALUT components are generated with gates, but can easily be replaced with full-custom autogenerated components (see Chapter 8). The circuit is pipelined to provide solutions every cycle, but based on modern synchronous embedded memory interfacing, the minimum achievable latency is four cycles. An alternative approach is to assume no clocked memory components (LUTs and RALUTs implemented with gates only), thereby lumping all the operations into a single clock cycle. This could lead to more switching, but it may be minimized with a good synthesis tool.

All software, HDL source files, and detailed results can be found at http://research.muscedere.com.

References

1. M.G. Arnold, T.A. Bailey, J.R. Cowles, J.J. Cupal, Redundant Logarithmic Arithmetic, *IEEE Transactions on Computers*, 39, 1077–1086, 1990.
2. E.E. Swartzlander Jr., A.G. Alexopoulos, The Sign/Logarithm Number System, *IEEE Transactions on Computers*, C-24, 1238–1242, 1975.

3. T. Stouraitis, V. Paliouras, Considering the Alternatives in Low-Power Design, *Circuits and Devices Magazine*, 17, 22–29, 2001.
4. J.N. Coleman, E.I. Chester, C.I. Softley, J. Kadlec, Arithmetic on the European Logarithmic Microprocessor, *IEEE Transactions on Computers*, 49, 702–715, 2000.
5. N.G. Kingsbury, P.J.W. Rayner, Digital Filtering Using Logarithmic Arithmetic, *Electronics Letters*, 7, 56–58, 1971.
6. D.M. Lewis, Interleaved Memory Function Interpolators with Application to an Accurate LNS Arithmetic Unit, *IEEE Transactions on Computers*, 43, 974–982, 1994.
7. D.M. Lewis, An Architecture for Addition and Subtraction of Long Word Length Numbers in the Logarithmic Number System, *IEEE Transactions on Computers*, 39, 1325–1336, 1990.
8. F.J. Taylor, R. Gill, J. Joseph, J. Radke, A 20 Bit Logarithmic Number System Processor, *IEEE Transactions on Computers*, 37, 190–200, 1988.
9. R. Muscedere, V. Dimitrov, G.A. Jullien, W.C. Miller, Efficient Techniques for Binary-to-Multidigit Multidimensional Logarithmic Number System Conversion Using Range-Addressable Look-Up Tables, *IEEE Transactions on Computers*, 54, 257–271, 2005.
10. T. Stouraitis, F.J. Taylor, Analysis of Logarithmic Number System Processors, *IEEE Transactions on Circuits and Systems*, 35, 519–527, 1988.
11. T.G. Robertazzi, S.C. Schwartz, Best "Ordering" for Floating-Point Addition, *ACM Transactions on Mathematics Software, 14*, 101–110, 1988.
12. F. de Dinechin, A. Tisserand, Multipartite Table Methods, *IEEE Transactions on Computers*, 54, 319–330, 2005.
13. J. Detrey, F. de Dinechin, A VHDL Library of LNS Operators, in *Conference on Record of the 37th Asilomar Signals, Systems and Computers*, 2003, vol. 2, pp. 2227–2231.
14. M.G. Arnold, C. Walter, Unrestricted Faithful Rounding Is Good Enough for Some LNS Applications, in *Proceedings of 15th IEEE Symposium on Computer Arithmetic*, 2001, pp. 237–246.

8

Optimizing MDLNS Implementations

8.1 Introduction

This chapter demonstrates a number of optimizations that can significantly improve the performance of MDLNS. In particular, we show how an optimal base can significantly reduce the range of the second base exponent and therefore reduce the hardware required. Using such optimizations allows our architectures to become much more competitive with existing systems based on fixed-point and floating point binary as well as those based on LNS. We also show that migrating from a two-bit signed system to a one-bit signed system can halve the computation time required to determine the optimal base, as well as reduce the critical paths of the architectures presented in earlier chapters. The main contents of this chapter are based on a 2008 publication by R. Muscedere [1].

8.2 Background

8.2.1 Single-Digit 2DLNS Representation

We start our discussion by examining the single-digit two-dimensional logarithmic number system (2DLNS) case:

$$x = s \cdot 2^a \cdot D^b + \varepsilon \tag{8.1}$$

where ε is the error in the representation. The sign, s, is typically –1 or 1, but the case $s = 0$ is also required when either the number of digits required to represent x is less than n, or the special case when $x = 0$. The second base, D, is our target for optimization. It should be chosen such that it is relatively prime to 2, but it does not necessarily need to be an integer, especially in signal processing applications. Allowing real numbers for D can greatly increase the chance of obtaining an extremely good representation (i.e., with very small exponents) for a given set of numbers, especially with two or more digits. The exponents are integers with a constrained precision. R is the bit width of the second base exponent, such that $b = \{-2^{R-1}, \ldots, 2^{R-1}-1\}$. This value directly affects the complexity of the MDLNS system. We shall also define B as the bit width of the binary exponent, such that $a = \{-2^{B-1}, \ldots, 2^{B-1}-1\}$. Later, when we look at a practical example, the resolution of these exponent ranges will be further refined since the full bit range can be rather large. Unlike R, the value of B does not directly affect the complexity of the system.

8.2.2 Single-Digit 2DLNS Inner Product Computational Unit

Figure 8.1 shows the architecture [2] of the single-digit 2DLNS inner product computation unit (CU), as earlier discussed in Chapter 5. To recap, the multiplication is performed by small parallel adders for each of the exponents of the operands (top of the figure). The output from the second base adder is the address for a lookup table (LUT) (ROM), which produces an equivalent floating point value for the product of the nonbinary bases

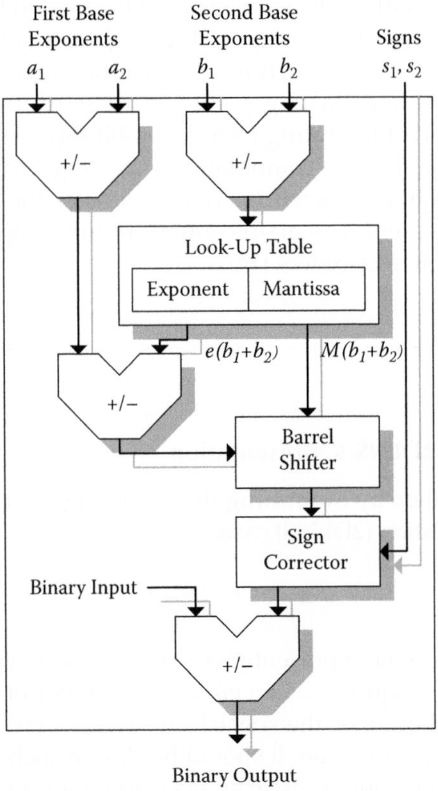

FIGURE 8.1
One-digit 2DLNS inner product computational unit. (From R. Muscedere, Improving 2D-Log-Number-System Representations by Use of an Optimal Base, *Eurasip Journal on Advances in Signal Processing*, 2008, 1–13, 2008. With permission.)

(i.e., $D^{b_1+b_2} \approx 2^{e(b_1+b_2)} \times M^{(b_1+b_2)}$). The base 2 exponents are added to those of the table to provide the appropriate correction to the subsequent barrel shifter (i.e., $2^{a_1+a_2} \times D^{b_1+b_2} \approx 2^{a_1+a_2+e(b_1+b_2)} \times M^{(b_1+b_2)}$). This result may then be converted to a two's complement representation, set to zero, or unmodified based on the product of the signs of the two inputs (–1, 0, or 1, respectively). The final result is then accumulated with prior results to form the total accumulation (i.e., $y(n+1) = y(n) + 2^{a_1+a_2+e(b_1+b_2)} \times M^{(b_1+b_2)}$).

This structure removes the difficult operation of addition/subtraction in 2DLNS by converting the product into binary for simpler accumulation. It is best for feed-forward architectures. We note that when the range of the second base exponent, R, of the 2DLNS representation is small (e.g., less than 4 bits), then the LUTs will also be very small because of the exponential relationship between LUT size and input address range.

The structure can be extended to handle more bases by concatenating the output of each corresponding exponent adder to generate the appropriate address for the LUT. The penalty, however, is that every extra address bit doubles the LUT entries. The structure itself will be replicated depending on the number of digits. If both operands have the same number of digits, we can expect to have n^2 such units in an n-digit MDLNS. For a parallel system, these outputs can be summed at the end of the array using, for example, an adder tree. The greatest advantage of the use of more than one digit for the operands is that one can obtain extremely accurate representations with very small second base exponents. However, the area cost increases as the number of computational channels increase.

8.3 Selecting an Optimal Base

8.3.1 The Impact of the Second Base on Hardware Cost

A closer look into the architecture above shows that the LUT stores the floating point-like representation of the powers of the second base, D. The area complexity depends almost entirely on the size of the LUT, which is determined by the range of the sum of the second base exponents, b_1 and b_2. Our main goal in selecting the second base is to reduce, as much as possible, the size of the largest second base exponents used while maintaining the application constraints. The actual value of D can be selected to optimize the implementation without changing the overall complexity of the architecture; in fact, as we shall see, such an optimization offers great potential for further reductions of the hardware complexity. Therefore, any value of D will only change the contents of the LUT, while the range of the second base exponents is the only factor that influences the size of the LUT. The same can be said for the binary-to-MDLNS converters covered in Chapter 6. Their complexity is limited by this range as well as the number of digits.

8.3.2 Defining a Finite Limit for the Second Base

We can limit the potential range of what could be considered to be an optimal value by analyzing the unsigned single-digit representation, as shown in Equation (8.2).

$$2^a D^b = 2^{a-b}(2D)^b = 2^{a+b}(D/2)^b \tag{8.2}$$

Equation (8.2) shows that we can multiply or divide the unknown base by any multiple of the first base, thereby changing its exponent but not changing the computational result. This simple relationship implies a restriction on the range of values of an optimal base. For example, if our search were to begin at $D = 3$, then it would be pointless to go outside of the range 3 to 6 because the results of the representation would simply be repeated.

The relationship in Equation (8.2) also demonstrates that as the value of D is divided by a multiple of 2, the exponent of the first base will increment by 1 when b is positive but decrease by 1 when b is negative. A similar conclusion can be made for the case when D is multiplied by a multiple of 2. Therefore, some representations may have large values for the first base exponent, and some may have smaller values. For a hardware implementation, the bit width of the first base exponent should be reduced while maintaining the selected representation space. We can determine the bit width for the first base exponent by limiting our representation range with Equation (8.3).

$$1 \le 2^a D^b < 2 \tag{8.3}$$

There is a unique first base exponent for every second base exponent. We continue by taking the logarithm of (8.3) as shown in (8.4).

$$0 \le a\ln(2) + b\ln(D) < \ln(2) \tag{8.4}$$

From (8.4) we obtain limits on the first base exponent, as shown in (8.5).

$$-b\frac{\ln(D)}{\ln(2)} \le a < 1 - b\frac{\ln(D)}{\ln(2)} \tag{8.5}$$

Since the range of b is known, the value of a can be found for all valid values of b. From this, the integer range of a can be found from the maximum and minimum values of b. The binary word length of the usable 2DLNS range is added to the maximum integer range of a to find the total range of a. For example, if $D = 3$ and b ranges from –4 to 3 (4 bits), then the range for the first base exponent will be between –4 and 7 for numbers between 1 and 2. If we wish to represent, at most, a 9-bit integer, then we will require a range of [–4, (7 + 9 = 16)] for the first base exponent, or 6 bits.

Using these relationships we can potentially reduce the number of bits required to represent a. From Equation (8.4), the range of a depends on the ratio $\ln(D)/\ln(2)$, where minimizing $\ln(D)$ results in a smaller bit width for a. We also note that the product of D with any power of 2 will produce the same 2DLNS results, as shown in Equation (8.6).

$$\frac{\ln(2^k D)}{\ln(2)} = k\frac{\ln(D)}{\ln(2)} \tag{8.6}$$

The function $\ln(D)$ will be minimized when D is closest to 1. The optimal range of D can thus be found by relating $\ln(y)$ (which is > 1) with $\ln(y/2)$ (which is < 1). Setting $\ln(y) = -\ln(y/2)$ we obtain $y = \sqrt{2}$. Therefore the optimal range of D is between $\sqrt{2}/2$ (or $1/\sqrt{2}$) and $\sqrt{2}$. We now have established an optimal range for D that will provide a minimal bit width to represent the first base exponent, a, and eliminate base replication.

If we rework our previous example using $D = 3/4 = 0.75$ and set the range of b to $[-4, 3]$ (4 bits), the range for the first base exponent will be between -1 and 2. To represent a maximum of a 9-bit integer, we will require a range of $[-1,(2 + 9 = 11)]$ for the first base exponent, or 5 bits. This is a savings of 1 bit from the previous example, where $D = 3$, but with no change in the representation.

8.3.3 Finding the Optimal Second Base

We have developed two methods for determining the optimal base for m numbers in the set x. The first, an algorithmic approach, only applies to single-digit 2DLNS, and the second, a range search, applies to any number of digits.

8.3.3.1 Algorithmic Search

Using the assumption that the optimal base represents one of the values in the given set, x, with virtually no error ($\varepsilon \cong 0$), then that optimal base can be found by solving for the base from the single-digit unsigned 2DLNS expression as shown in Equation (8.7):

$$D = \sqrt[b]{\frac{x}{2^a}} = x^{1/b}2^{-a/b} \tag{8.7}$$

This expression can be solved for every value in the set x given the range of b that depends on R (i.e., $b = \{-2^{R-1}, \ldots, -1, \ldots, 2^{R-1}-1\}$). Since any multiple of 2^k on D does not affect the 2DLNS representation, a is limited by b, such that $a = \{-b + 1, \ldots, b - 1\}$. Although many solutions may exist depending on the value of R and the number of values x, only the bases with the smallest errors will be finely adjusted until the final optimal base is found (see below).

8.3.3.2 Range Search

A second alternative is to perform a range search through all the possible real bases. We have already seen that the most efficient bases for hardware implementation lie in the range $\left[1/\sqrt{2},\sqrt{2}\right]$. This limitation offers a practical start and end point for a range search. Given an arbitrary second base, we have written a program that measures the error of mapping the given set x into a multidigit 2DLNS representation. The possible representation methods can reflect those of hardware methods available, such as the greedy/quick, high/low approximations or a brute-force approach. Our program uses a dynamic step size that is continuously adjusted by analyzing the change in the mapping errors for a series of test points. This step size increases so long as the resulting errors are monotonically improving. If this is not the case the program retraces and decreases this step size. When a better error is found it is added to a running list of optimal candidates. Using a dynamic step size is effective in finding optimal base candidates while also reducing the overall search time. Once the entire range has been processed, each element in this list is finely adjusted. Depending on the representation method selected and the range of R, this approach can generate fewer bases than the algorithmic method and therefore produce results in a shorted amount of time.

8.3.3.3 Fine Adjustment

A fine adjustment can be performed with the list of optimal candidates by progressively adding and subtracting smaller and smaller values. The performance of the software is further increased by using direct floating point (IEEE 64-bit) manipulation as well as minimizing conditional branches and expensive function calls. This approach drastically improves search times by initially performing a coarse search, by one of the methods above, and then a finer search near the selected optimum points.

8.4 One-Bit Sign Architecture

The data path of the 2DLNS processor (in Figure 8.1) is affected significantly by the signs of the operands. The required sign correction operation comes at a cost of additional logic and power. Thus far, a multidigit architecture would require additional processing to be performed after the 2DLNS processor, such as summing all the channels. It is possible to use the common one-bit sign binary representation for the intermediate results. We have therefore developed a new 2DLNS sign system to reduce the processing path of the 2DLNS inner product computational unit (CU) while producing a single sign bit binary representation.

8.4.1 Representation Efficiency

Our original 2DLNS notation uses two bits to represent the sign for each digit (–1, 0, and 1); however, only three of four states are used, one of which (zero) only represents a single value. By using two bits for the sign, the efficiency of the representation is approximately 50%.

$$efficiency_{two-bit\,sign} = \frac{\text{valid representations}}{\text{total possibilities}}$$

$$= \frac{2^{1+B+R}+1}{2^{2+B+R}} \cong 0.5$$

To improve this efficiency we propose that only a single sign bit is needed to represent the most common cases, that is, –1 and 1. We then choose to represent zero by setting the second base exponents to their most negative values (i.e., if the range is [–4, 3], then –4 is used to represent zero). This allows us to reduce the circuitry of the system while maintaining the independent processing paths of the exponents; this modification is easily integrated into the existing two-bit sign architecture. This special case for zero still leaves us with a significantly smaller unused representation space compared to the two-bit sign system. As R increases, the efficiency of the representation approaches 100%.

$$efficiency_{one-bit\,sign} = 1 - \frac{\text{invalid representations}}{\text{total possibilities}}$$

$$= 1 - \frac{2^{1+B}}{2^{1+B+R}} = \frac{2^R - 1}{2^R} \to 1$$

With the one-bit sign system, the range of the second base changes to $b_i = \{-2^{R-1}+1, \ldots, 2^{R-1}-1\}$ with a special case of $b_i = -2^{R-1}$ representing zero.

8.4.2 Effects on Determining the Optimal Base

Since the upper and lower bounds of the second base exponent are equal in magnitude, this eliminates the need for any reciprocal computations in determining the optimal base (i.e., $D^b = 1/D^{-b}$), thus approximately halving the search time for both algorithms. For the algorithmic search, the possible range of b is changed, such that $b = \{1, \ldots, 2^{R-1}-1\}$. For the range search approach the second base limits are now $[1/\sqrt{2}, 1.0]$ or $[1.0, \sqrt{2}]$.

8.4.3 Effects on Hardware Architecture

By using the one sign bit architecture, the word length for any 2DLNS representation is reduced by 1 bit per digit. Compared to the original CU, we can remove the sign corrector component (essentially a conditional two's complement generator). The sign is calculated by simply XORing the two signs of the inputs. The output is now in a binary representation that can easily be manipulated further with the sign bit depending on the number of digits (see Figure 8.2). The special zero case only need be handled by modifying the very small adders in the multiplication component; the representation of zero is now inside the table, which therefore eliminates the conditional path.

To accumulate this result with any other value, we can use the generated sign bit to determine the proper operation of an addition/subtraction component (see Figure 8.3). The inclusion of a one-bit sign allows us to reduce the hardware and computational path by removing the zero/two's complement generator. The final adder/subtractor component is slightly larger than an adder, but with regards to the whole system, this architecture will consume less area and time.

In the case of a two-digit 2DLNS system, the accumulation of the four output channels can be simplified with the one-bit sign by using only three adder/subtractor components and simple logic to coordinate the proper series of operations (see Figure 8.4). The processing delay from the LUTs is only three arithmetic operations, and the overall logic is also reduced since the three adder/subtractor components are smaller than the three separate adders and four two's complement generator components present in the original CU. This approach was used in [3] and demonstrated a 55% savings in power, as well as other improvements, compared to the original design in [4].

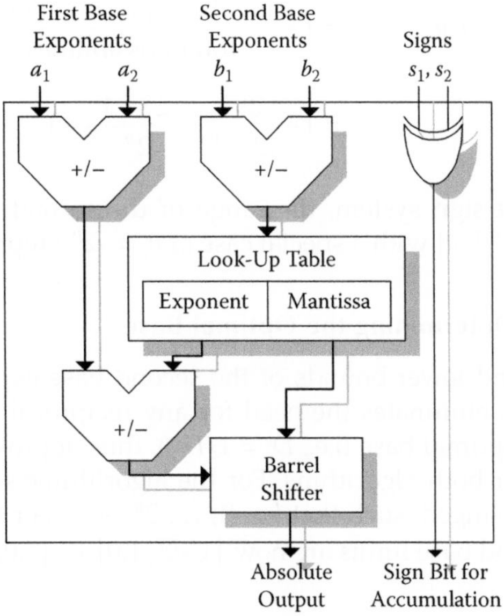

FIGURE 8.2
One-bit sign 2DLNS inner product computational unit. (From R. Muscedere, Improving 2D-Log-Number-System Representations by Use of an Optimal Base, *Eurasip Journal on Advances in Signal Processing*, 2008, 1–13, 2008. With permission.)

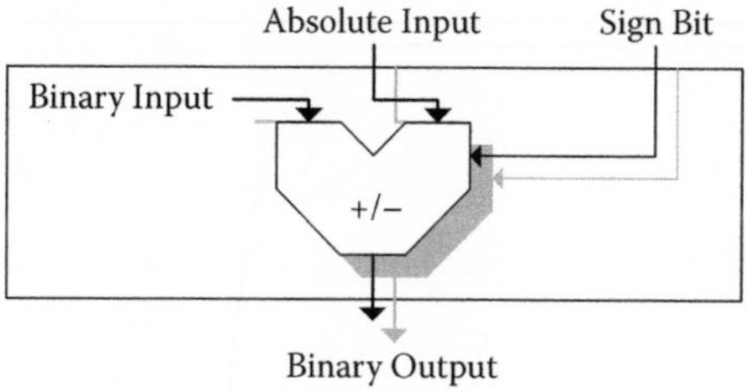

FIGURE 8.3
Single-digit one-bit sign accumulation component. (From R. Muscedere, Improving 2D-Log-Number-System Representations by Use of an Optimal Base, *Eurasip Journal on Advances in Signal Processing*, 2008, 1–13, 2008. With permission.)

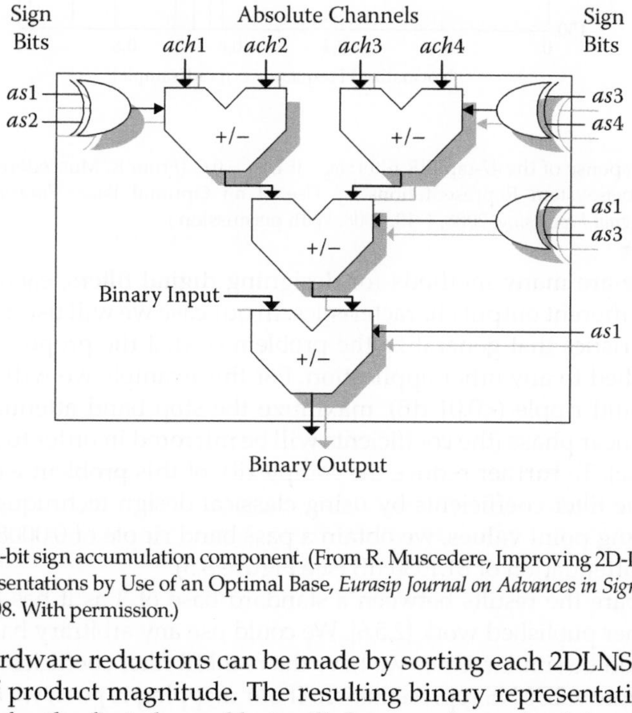

FIGURE 8.4
Two-digit one-bit sign accumulation component. (From R. Muscedere, Improving 2D-Log-Number-System Representations by Use of an Optimal Base, *Eurasip Journal on Advances in Signal Processing*, 2008, 1–13, 2008. With permission.)

Further hardware reductions can be made by sorting each 2DLNS processor in order of product magnitude. The resulting binary representation will be the largest for the first channel but will decrease for each of the subsequent channels. If the range of both operands is known, the mantissa in the LUTs can be sized correctly as well as the subsequent adders.

8.5 Example Finite Impulse Response Filter

To demonstrate how important it is to choose an optimum base, D, we provide the following example of a 47-tap low-pass finite impulse response (FIR)

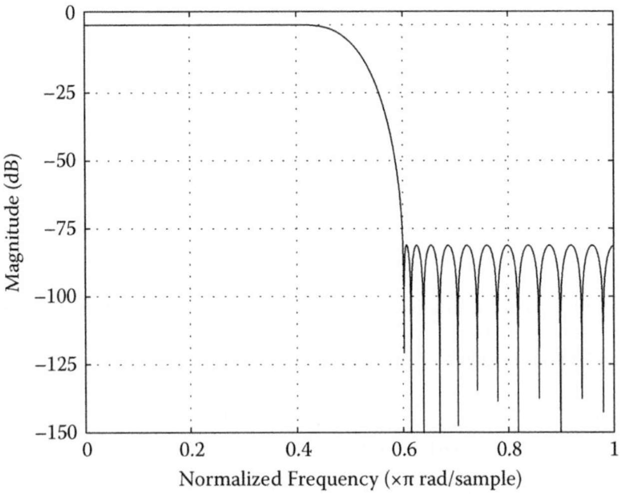

FIGURE 8.5
Magnitude response of the 47-tap FIR filter; $\omega_p = 0.4$, $\omega_s = 0.6$. (From R. Muscedere, Improving 2D-Log-Number-System Representations by Use of an Optimal Base, *Eurasip Journal on Advances in Signal Processing*, 2008, 1–13, 2008. With permission.)

filter. There are many methods for designing digital filters, each of which prioritize different output characteristics. In our case we will use a simple set of characteristics that generalize the problem so that the proposed method can be applied to any other application. For this example we will minimize the pass band ripple (<0.01 dB), maximize the stop band attenuation, and maintain linear phase (the coefficients will be mirrored in order to guarantee linear phase). To further reduce the complexity of this problem we will first generate the filter coefficients by using classical design techniques. Ideally, using floating point values, we obtain a pass band ripple of 0.0008 dB and a stop band attenuation of 81.1030 dB (see Figure 8.5).

We compare the results between a standard base of 3 as it has been used often in other published work [2,5,6]. We could use any arbitrary base and the results would be similar. Once we have the real FIR filter coefficients, we then map them into a 2DLNS representation. If our mapping is poor, we can expect both poor stop band attenuation as well as pass band ripple, whereas a more accurate mapping will result in better filter performance. We choose not to calculate the filter's performance during the calculation of the optimal base, but rather the absolute error in the mapping itself. This improves the performance of the optimal base calculation and allows the process to be used with any filter design technique or even for other entirely different applications. Note that we do not impose restrictions on the size of the binary exponents as they contribute very little to the overall complexity of the architecture. We require, however, to know what their range will be for hardware implementation. As discussed in earlier chapters, a FIR filter computes inner products between

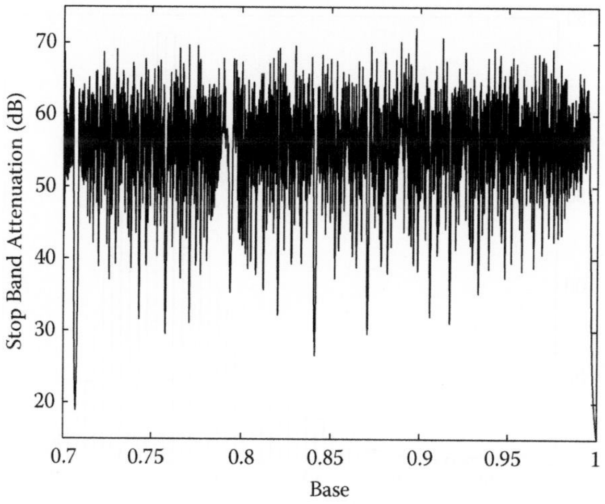

FIGURE 8.6
Stop band attenuation for bases 0.7071 to 1.0000 for $r_c = 63$; worst case is 15.0285dB, best case is 72.0858 dB. (From R. Muscedere, Improving 2D-Log-Number-System Representations by Use of an Optimal Base, *Eurasip Journal on Advances in Signal Processing*, 2008, 1–13, 2008. With permission.)

input data (from some external source) and a set of filter coefficients to generate output data. Since we are discussing the 2DLNS representations of the data, coefficient, and output values in the same system, we shall refer to their exponents as (a_d, b_d), (a_c, b_c), and (a_o, b_o), respectively. We will also compensate for a finer resolution on the multiplicands such that $b_d = \{-r_d, \ldots, r_d\}$ and $b_c = \{-r_c, \ldots, r_c\}$, resulting in a product where $r_c = r_d + r_c$ and $b_c = \{-r_c, \ldots, r_c\}$. The range of the second base exponent, b_o, will dictate the complexity of the system.

To demonstrate how the selection of any arbitrary base can affect the filter performance, we have mapped the coefficients into a single-digit 2DLNS using bases between $1/\sqrt{2}$ and 1.0 (in increments of 0.0001) and plotted the resulting stop band attenuation in Figure 8.6 for $r_c = 63$.

This figure clearly shows that there is no obvious correlation between the filter's performance and the choice of the second base; in fact, it appears random. The same can be said for the pass band ripple. We can also examine these results in the form of a histogram, as in Figure 8.7.

The low values of the stop band attenuation are a result of bases very close to 1.0 (where, as the exponent increases, the normalized representation approaches 1). The average is 54.6721 dB, but for a base of 3 (or 0.75) it is 61.0460 dB, which is better than the average results. Even though our sample size for this test is small (2,930 values), it is reasonable to assume that any arbitrary base will not give the best 2DLNS representational performance. The best base for this set of data is 0.8974 with a stop band attenuation of 72.0858 dB. This is a good result, but it is possible to achieve even better without testing the performance of the filter for every possible base.

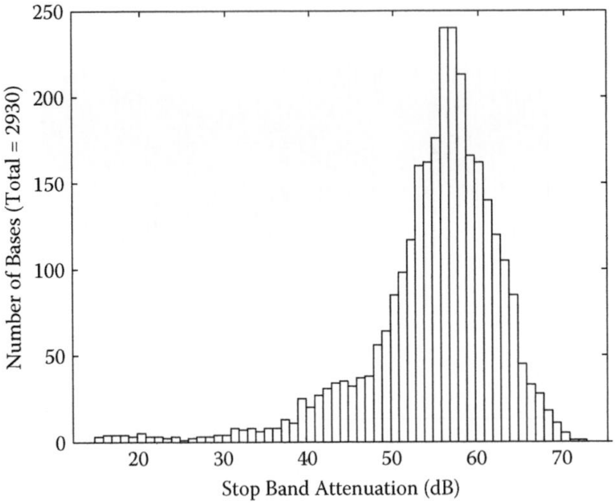

FIGURE 8.7
Histogram of stop band attenuation for bases 0.7071 to 1.0000. (From R. Muscedere, Improving 2D-Log-Number-System Representations by Use of an Optimal Base, *Eurasip Journal on Advances in Signal Processing*, 2008, 1–13, 2008. With permission.)

8.5.1 Optimizing the Base through Analysis of the Coefficients

Generally the input data are large in magnitude and, in order to accommodate for this, we will need to use two or more digits for their representation. If the input data were relatively small we could use a one-digit representation; however, we would expect some quantization error. For our example we use a two-digit representation, as the intended input range is larger (–32,768 to 32,767 or 16 bits).

8.5.1.1 Single-Digit Coefficients

In Chapter 2, we demonstrated that the coefficient distribution of many different types of FIR filters is better represented by a logarithmic-like number system (such as the LNS or 2DLNS) rather than a linear number representation (such as binary). Therefore we should be able to obtain very good single-digit approximations in the 2DLNS by making use of a carefully calculated second base. Since the data representation uses two digits, the resulting system will consist of only two computational channels. Later in this chapter we will also consider a two-digit 2DLNS representation with four channels.

A comparison of the frequency response for a wide range of exponent ranges (or various values of r_c) for the example filter is shown in Table 8.1. We compare the pass band ripple, stop band attenuation between a system with the second base of 3 and an optimal base.

Table 8.1 demonstrates that as r_c increases we can save up to two bits on the second base exponent by using an optimal second base rather than 3.

TABLE 8.1

Filter Performance for Ternary and Optimal Base (Single Digit)

	Base 3		Optimal Base		
r_c	Pass Band Ripple (dB)	Stop Band Atten. (dB)	Pass Band Ripple (dB)	Stop Band Atten. (dB)	Base
1	0.5147	24.2840	0.1498	35.3497	0.793123
3	0.2644	30.1926	0.0518	44.5401	0.889553
7	0.0526	44.3046	0.0202	52.5938	0.888477
15	0.0195	52.8771	0.0053	64.4114	0.897501
31	0.0131	56.5177	0.0031	68.5123	0.897504
63	0.0079	61.0460	0.0014	74.6231	0.788051
127	0.0034	68.7088	0.0010	77.6842	0.908324
255	0.0020	73.4312	0.0010	80.0042	0.876152

The size of the second base exponent plays an important role in the size of the hardware due to the required LUTs; the LUT size for any nonbinary base grows exponentially with increase in the nonbinary exponent, whereas an increase in the binary exponent adds minimal hardware. Any change to the second base, including using real numbers rather than integers, has no impact on the cost of the hardware. Therefore hardware designed for a second base of 3 is easily converted to use the optimal base since we are only changing the contents of the tables and not their dimensions. In this case a two-bit reduction translates to a 4 times area savings per LUT or CU.

The optimal base is truncated to six decimal digits for presentation; however, the number of decimal digits is computed up to 15 (IEEE 64-bit floating point), and this accuracy may be very necessary when the exponents on the base are large.

8.5.1.2 Two-Digit Coefficients

Here we use a two-digit representation for both the signal and the coefficients. This will result in four parallel computational units. The method for generating these representations is via a *brute-force* approach where all possible representations are generated and the one with the lowest error is chosen. This method is not applicable to hardware implementation since it is assumed that the coefficients will be generated offline. This approach was taken in [5] to improve eight separate FIR filters in a filterbank application. A comparison of the frequency response for various values of r_c is shown for the two-digit coefficients in Table 8.2.

We stop at $r_c = 7$ since the results are approaching near ideal. Again, the use of an optimal second base offers the same stop band attenuation as with a second base of 3 but with two fewer bits. This is an important saving since the CU is duplicated four times.

TABLE 8.2

Filter Performance for Ternary and Optimal Base (Two Digit)

	Base 3		Optimal Base		
r_c	Pass Band Ripple (dB)	Stop Band Atten. (dB)	Pass Band Ripple (dB)	Stop Band Atten. (dB)	Base
1	0.0314	48.8313	0.0013	76.8648	0.757910
2	0.0042	66.4051	0.0010	78.6542	0.899757
3	0.0031	68.9081	0.0009	80.0583	0.957323
4	0.0020	73.0290	0.0008	80.3113	0.789982
5	0.0016	74.9437	0.0008	80.9408	0.814313
6	0.0016	74.7669	0.0008	81.0408	0.919486
7	0.0015	75.4867	0.0008	81.0342	0.885619

8.5.1.3 Comparison of Single- and Two-Digit Coefficients

In order for a one-digit 2DLNS to achieve 80 dB stop band attenuation, we need to use 9 bits ($r_c = 255$) for the second base exponents, and correspondingly, we require a LUT with 512 entries for each CU (two for a parallel implementation). For a two-digit 2DLNS we only need 3 bits ($r_c = 3$) to represent the second base exponents, therefore requiring a LUT with eight entries for each CU (four CUs are required for a parallel implementation). The two-digit coefficient system appears to be favorable since the LUTs are smaller; however, there is some additional overhead in the accumulation circuit for all the channels. It is also very important to note that this entire four-channel architecture is multiplier-free since it consists only of small adders and very small LUTs.

8.5.1.4 Effects on the Two-Digit Data

Clearly the choice of the second base has a significant effect on the performance of the filter. However, in order to use this representation effectively, we also have to apply the same second base to the data representation in order for the 2DLNS arithmetic to operate properly. In the case of filter design the optimal base is selected by the performance of the filter, which we can relate back to the quality of the mapping. In the case of data, specifically integers, we do not necessarily require perfect mapping but only representations where $\varepsilon \le 0.5$ (see Chapter 5) [7]. Table 8.3 shows the range of r_d for 0%, 1%, 5%, and 10% non-error-free representations with a base of 3 and the optimal bases from Tables 8.1 and 8.2, respectively.

We can see that the optimal base for the best filter performance is not ideal for data mapping since r_d, on average, must be in the hundreds. When applying the optimal base to the coefficients, the performance increases as

TABLE 8.3

Data Representation Performance for Various Bases

	Error-Free	1% Non-Error-Free		5% Non-Error-Free		10% Non-Error-Free	
Base	r_d	r_d	**Worst ε**	r_d	**Worst ε**	r_d	**Worst Dev.**
3.000000	93	51	0.898	28	1.272	24	2.533
0.793123	147	131	1.867	113	17.349	101	28.392
0.889553	71	36	1.213	32	4.452	30	5.976
0.888477	70	48	1.002	35	1.872	27	2.009
0.897501	59	44	1.290	34	2.435	28	2.438
0.897504	54	45	1.330	35	2.437	29	2.522
0.788051	105	74	1.189	55	1.858	42	2.308
0.908324	66	47	1.364	36	1.720	29	2.121
0.876152	74	59	1.511	46	2.763	39	4.009
0.757910	886	734	6.002	566	18.941	496	25.501
0.899757	52	40	0.994	28	1.605	24	2.476
0.957323	67	54	1.810	45	3.238	39	3.678
0.789982	70	45	1.235	29	1.414	23	2.565
0.814313	270	191	1.252	134	2.206	106	2.944
0.919486	95	70	1.106	50	1.707	39	2.636
0.885619	48	40	1.442	30	1.545	25	2.751

r_c increases. There is no correlation here as the base was chosen only for optimal mapping of the coefficients. The case where $r_d = 886$, in particular, is unusual, as this base produces bit streams with long sequences of ones or zeros when the exponent exceeds 1.

8.5.2 Optimizing the Base through Analysis of the Data

We have seen how applying an optimal base to the coefficients of a digital filter can significantly increase the accuracy of the 2DLNS representation. This same improvement can be seen when applied to the input data of the filter. For the case of a 16-bit signed input (with a range [−32768, 32767]), we require that $r_d = 39$ in order to achieve a completely error-free mapping using the high/low method (see chapter 6). For particular applications, however, a complete error-free mapping may not be necessary. Table 8.4 summarizes different choices of r_d for non-error-free integer mappings.

The trend of the number of non-error-free representations follows an exponential decay as r_d increases. From the optimal base calculations of the coefficients (Table 8.3), we have the smallest r_d of 36 with 1% non-error-free representation but with a worst-case error of 4.452. The next smallest r_d of 40 offers a worst-case error of only 0.994. Both cases require r_d to

TABLE 8.4

Data Representation Performance for Optimal Bases

Base	r_d	Non-Error-Free Reports	Worst ε	% Non-Error-Free
0.810537	16	16,472	2.933	25.13
0.853451	17	14,296	2.857	21.81
0.844819	18	12,406	2.362	18.93
0.867483	19	10,926	2.156	16.67
0.776015	20	9,238	1.991	14.10
0.797969	21	7,836	1.717	11.96
0.769616	22	6,630	1.598	10.12
0.915321	23	5,566	1.540	8.49
0.855890	24	4,594	1.333	7.01
0.838039	25	3,746	1.343	5.72
0.987020	26	3,020	1.231	4.61
0.797037	27	2,398	1.104	3.66
0.843455	28	1,854	1.061	2.83
0.719670	29	1,394	1.001	2.13
0.815027	30	1,082	0.906	1.65
0.785026	31	766	0.845	1.17
0.749814	32	558	0.772	0.85
0.710762	33	348	0.822	0.53
0.843007	34	216	0.712	0.33
0.990287	35	144	0.684	0.22
0.892307	36	74	0.636	0.11
0.854608	37	36	0.610	0.05
0.756487	38	16	0.569	0.02
0.735582	39	0	—	0.00

be increased by more than 33% to achieve an error-free representation. When optimizing the base for the data representation, we can select $r_d = 32$ to achieve less than 1% non-error-free representation with a worst-case $\varepsilon = 0.772$. This is comparable to $r_d = 40$ in Table 8.3 but with a 25% reduction in the exponent range as well as the LUT entries. This approach was used in [3] in order that the filter coefficients could be changed by mapping them into the optimal base selected for the data representation. This required a larger r_c to improve the filter performance, but allowed the coefficients to be run time loaded.

8.5.3 Optimizing the Base through Analysis of Both the Coefficients and Data

We have so far seen that an optimal base can improve the coefficient or data representations of a 2DLNS filter architecture without changing the range of the exponents. Again, the 2DLNS arithmetic will not operate correctly unless both bases are the same. In each case the selection of one base severely impacts the representation efficiency of the other. To remedy this situation, we have modified the optimal base software to target two separate scenarios. This is done by optimizing the two independent sets of values and minimizing the product of their errors.

8.5.3.1 Single-Digit Coefficients and Two-Digit Data

For our example of a FIR filter, the data are represented with two-digit 2DLNS (using the high/low method) and the coefficients with a single digit (later, two-digit brute-force method). Since the range of r_c must be large for the single-digit coefficients to obtain good filter performance, we will also target an error-free data mapping since we can expect that r_d will be close to 39. Through experimenting with different variations of r_d, it was found that $r_d = 42$ produced an error-free data representation. To maximize the data path utilization for r_o, the remaining bits are used to specify r_c; this technique has been used in almost all DBNS/MDLNS publications to date. Table 8.5 shows the best results of the optimal base calculations for 8 (42 + 85 = 127) and 9 (42 + 213 = 255) bits. Using these optimal bases produces a pass band ripple below the specification of 0.01 dB. The bold values on Table 8.5 indicate the best result for the selected attribute.

8.5.3.2 Comparison to the Individual Optimal Base

Comparing the filter performance results of Tables 8.1 to 8.5, we can see approximately a 2 dB reduction in the stop band attenuation. However, comparing the non-error-free data mapping to Table 8.3 we can see a large improvement in the representation. This improvement seems to justify the sacrifice of 2 dB in the stop band.

When considering a hardware implementation, r_o will never exceed ±255 for the 9-bit system. The 2DLNS-to-binary conversion tables will require $2r_o + 2$ entries, one of which is for the zero representation. We will therefore

TABLE 8.5

Combined Optimal Base (Single-Digit Coefficient, Two-Digit Data)

r_d	r_c	Base	Stop Band Atten. (dB)	Non-Error-Free Reps.	Worst ε	% Non-Error-Free
42	85	0.924440	**75.5074**	348	0.993	0.53
42	85	0.871988	68.7626	**0**	—	0.00
42	85	**0.815959**	75.1871	88	0.756	0.13
42	213	0.912396	**78.2265**	196	0.839	0.30
42	213	**0.872018**	77.2010	**0**	—	0.00

have two inner product CUs, each containing tables of 512 entries totaling 1,024 entries for both CUs.

8.5.3.3 Two-Digit Coefficients and Two-Digit Data

We can also apply the blended optimal base to the two-digit coefficient representation. Since the ranges on r_c are much smaller, we will explore the possibility of having a non-error-free data representation. As shown before, obtaining an error-free data representation will require larger ranges of r_d, which in turn will require larger tables for the four parallel inner product CUs. Table 8.6 shows various possibilities for r_d and r_c.

Initially 28 and 3 are chosen to maximize the bit width of the product exponent b_o, but the data representation is poor when the filter performance is high. As we increment r_c we can see an increase of about 0.5 dB for the best-case stop band attenuation. We settle on $r_c = 5$ as the best-case stop band of approximately 80 dB. As we increment r_d we see a similar exponential decay trend as before, when only optimizing for the data. In the case of maximum stop band attenuation, the number of non-error-free representations is quite high. This drops considerably when we sacrifice a little in the stop band (~0.1 dB). We can begin to reach an error-free data representation when $r_d = 40$ and above. Depending on the application, a non-error-free mapping may be acceptable considering that the worst-case ε is below 1.0.

8.5.3.4 Comparison to the Individual Optimal Base

When we compare the above results with the previous individual optimal bases, we can see that we have not sacrificed much in terms of stop band attenuation ($r_c = 5$) or exponent ranges for error-free data mapping ($r_d = 40$). This approach seems to offer the best filter performance and data representation as compared to the single-digit coefficients.

For the purposes of implementation, r_o will never exceed ±45. We can therefore expect to have four inner product CUs, each with 92 entries, totaling 368 entries for four CUs.

8.5.4 Comparison of Base 3 to the Optimal Bases

There are many possibilities available for an optimal base depending on the accuracy required for the filter performance and data representation. Table 8.7 compares the original base 3 and optimal base performance to provide at least 73 dB stop band attenuation and a 0% and 1% non-error-free data mapping. For all cases, the optimal base offers savings in the CU LUTs as well as for the range of the second base exponent.

In the single-digit case, we can increase or decrease r_d to decrease or increase the non-error-free representations, respectively.

TABLE 8.6

Combined Optimal Base (Two-Digit Coefficient, Two-Digit Data)

r_d	r_c	Base	Stop Band Atten. (dB)	Non-Error-Free Reps.	Worst ε	% Non-Error-Free
28	3	0.895008	**79.9529**	23,910	4.972	36.48
28	3	0.941001	66.9800	**1,908**	**1.000**	**2.91**
28	3	**0.737860**	77.5315	2,048	1.115	3.13
30	4	0.858929	**80.4562**	4,794	1.924	7.32
30	4	0.796725	71.2290	**1,062**	0.932	1.62
30	4	0.862663	75.2993	1,116	**0.875**	**1.70**
30	4	**0.745462**	80.3739	1,636	1.162	2.50
31	4	0.858929	**80.4562**	4,398	1.924	6.71
31	4	0.711017	69.9018	**828**	**0.824**	**1.26**
31	4	**0.745462**	80.3739	1,534	1.162	2.34
33	5	0.814313	**80.9394**	19,074	3.901	29.10
33	5	0.764157	75.2038	**394**	0.802	0.60
33	5	0.989748	75.1037	426	**0.738**	**0.65**
33	5	**0.816596**	80.5589	568	0.854	0.87
34	4	0.858929	**80.4563**	3,348	1.571	5.11
34	4	0.777919	72.8959	**246**	0.715	0.38
34	4	0.777915	75.5193	284	**0.693**	**0.43**
34	4	**0.790987**	79.2613	334	0.741	0.51
39	5	0.738251	**80.9350**	10,806	5.512	16.49
39	5	0.710235	76.9878	**4**	0.536	**0.01**
39	5	0.722788	75.8355	12	**0.516**	0.02
39	5	**0.732693**	80.4818	24	0.570	0.04
40	5	0.816596	**80.5634**	532	0.842	0.81
40	5	0.710131	72.6561	**0**	—	**0.00**
40	5	**0.974939**	80.2883	72	0.781	0.11
41	5	0.915757	**80.6101**	738	0.854	1.13
41	5	0.746550	77.9541	**0**	—	**0.00**
41	5	**0.938171**	80.0786	36	0.615	0.05
42	5	**0.974939**	**80.4183**	12	0.588	0.02
42	5	0.944509	78.6000	**0**	—	**0.00**

8.5.5 Comparison to General Number Systems

We have thus far only shown the improvement in the 2DLNS representation and circuit resources when applying an optimal base compared to the legacy base of 3. We can further compare the above results with those from common general number systems, such as fixed-point and floating point binary as well as a fixed-point exponent LNS, which are traditionally used in physical implementations. Table 8.8 shows a summary of the example

TABLE 8.7

Comparison of Standard and Optimal Base for a Minimum 73 dB System

Base	Stop Band Atten. (dB)	% Non-Error-Free	r_d	r_c	LUT Entries per CU	Total LUT Entries
Single-Digit Coefficients						
3.000000	73.4312	1.00	255	51	614	1,228
0.828348	73.7272	0.67	85	39	250	500
Savings					59.3%	
3.000000	73.4312	0.00	255	93	698	1,396
0.828348	73.7272	0.00	85	45	262	524
Savings					62.5%	
Two-Digit Coefficients						
3.000000	75.4867	1.00	7	51	118	472
0.777915	75.5193	0.43	4	34	78	312
Savings					33.9%	
3.000000	75.4867	0.00	7	93	202	808
0.746550	77.9541	0.00	5	41	94	376
Savings					53.5%	

filter's performance using these number systems for 1 to 20 bits. Note that the bit limitation is applied to the fractional portion of the representations only. The integer portion (i.e., exponent for floating point, integer for LNS exponent) is not considered since it only affects the fixed-point binary representation for values greater than 1; all of the filter coefficients are, however, less than 1.

From Table 8.6, we achieve greater than 80 dB stop band attenuation with only 6.6 bits per digit using a two-digit data/coefficient system. In this case, only 12 of the 65,536 inputs have any representation error, but it is less than 0.588. The LNS offers the best filter performance with 14 or more bits in the fractional exponent; however, this choice may come at the expense of a larger circuit for performing binary-to-LNS-to-binary conversion with 14 fractional bits as well as native LNS addition/subtraction, as in [8], for example.

In terms of the data representation, all of the above-mentioned general number systems will require at least 14 bits to represent an input integer with $\varepsilon < 1$. For this example, Table 8.7 shows that no more than 6.3 bits (per digit) is required when using a two-digit 2DLNS representation ($\varepsilon \leq 0.67$). For each of the general number systems considered, additional bits in the multipliers (or adders for LNS) will be required to include the calculation of the input data. The 2DLNS approach, presented in this chapter, does not need additional precision, but the two-digit system does require an accumulation stage to merge the results from the four separate processing channels.

TABLE 8.8

Comparison of Filter Performance for General Number Systems

	Fixed-Point Binary		Floating Point Binary		LNS with Fixed-Point Exponent	
Bits	**Pass Band Ripple (dB)**	**Stop Band Atten. (dB)**	**Pass Band Ripple (dB)**	**Stop Band Atten. (dB)**	**Pass Band Ripple (dB)**	**Stop Band Atten. (dB)**
1	6.0206	6.0206	1.4225	16.4174	2.2454	14.6224
2	3.6810	9.2326	1.2905	17.1989	1.3542	26.0100
3	3.6810	9.2326	0.1589	34.8402	0.5313	28.6544
4	1.9997	13.7377	0.1521	35.2187	0.3403	37.1995
5	0.5606	24.0824	0.0813	40.6289	0.1493	38.1408
6	0.2873	29.7543	0.0693	42.0232	0.0776	45.6083
7	0.2474	31.0320	0.0092	59.6780	0.0422	54.1982
8	0.0778	41.0001	0.0036	68.0561	0.0249	69.0545
9	0.0507	44.6930	0.0025	71.3190	0.0127	71.2978
10	0.0211	52.2642	0.0026	70.9529	0.0064	74.0112
11	0.0117	57.4244	0.0019	74.1708	0.0032	74.8158
12	0.0062	62.9227	0.0017	75.2752	0.0017	74.7228
13	0.0040	66.8129	0.0016	75.5764	0.0012	78.0202
14	0.0023	71.6093	0.0012	78.7328	0.0009	80.0245
15	0.0013	76.5792	0.0009	80.9136	0.0008	80.8327
16	0.0012	77.2776	0.0009	80.6924	0.0008	80.9733
17	0.0010	78.6727	0.0008	81.0919	0.0008	81.0450
18	0.0008	80.2013	0.0008	81.0246	0.0008	81.1399
19	0.0008	80.7061	0.0008	81.1587	0.0008	81.1113
20	0.0008	81.1363	0.0008	81.1301	0.0008	81.1117

The improvements to 2DLNS with an optimal base as well as a one-bit sign target applications where traditional number systems such as fixed-point and floating point binary as well as LNS are used. It will be interesting to see, in future research, what other applications see advantages in applying an optimized 2DLNS arithmetic.

8.6 Extending the Optimal Base to Three Bases

Applying our existing software to find a set of optimal bases on a multiple-base system would have serious scalability issues since we would be performing a linear search geometrically for each base. The complexity of this problem can be considered to be $O(n^k)$, where n is the number of scenarios tested for a single base, and k is the number of nonbinary bases. For the two-digit coefficient case of $r_d = 40$ and $r_c = 5$, n is 20,864,199. The computational time for any scenario varies considerably with r_d, r_c, and the representation method selected, since they all affect the data generation and search times. We do believe that it is possible to reduce this massive search time for a single-digit 3DLNS system (two nonbinary bases) by merging both algorithmic and range searching methods. By assuming that one of the target representations will have a near-zero error, the algorithmic search should provide a series of second bases with the range search supplying the third base. We have yet to attempt this process, but we believe it can improve the single-digit representation by using multiple bases with smaller exponent ranges rather than a single base with a larger exponent range. The complexity should be reduced as we are effectively performing many two-base optimizations (potentially in parallel) but with a much smaller limitation on the exponents. Using a 3DLNS system with smaller values of r could result in a more compact hardware implementation and better representation than a 2DLNS system.

8.7 Summary

Since the 2DLNS inner product computational unit requires a LUT for 2DLNS-to-binary conversions, it is important to try to minimize the size of this LUT because it is used many times in the overall system. We have shown, by example, that selecting an optimal second base can reduce the size of these LUTs and, more importantly, the range of b_o, by a minimum of 33% in one case without impacting the quality of the 2DLNS representation compared to the standard second base of 3. These results should hold true for any other arbitrary second base. As the binary-to-2DLNS conversion hardware also depends on the range of the second base exponent, selecting an optimal base will offer reductions in this component and may potentially reduce the data paths and storage registers of the complete system. Reducing the range of b_o can also impact other architectures, such as the native MDLNS addition/subtraction circuit shown in Chapter 7.

By migrating from a two-bit sign to a one-bit sign, the computation time of the optimal base halves and the hardware area and critical paths of published architectures are further reduced as some processing steps are eliminated.

The software for finding an optimal base can be utilized in many different applications other than FIR filter design. The software accepts either a list of real numbers, a range of integers, or both in order to find the best representation with minimal error.

All software, source files, and detailed results can be found at http://research.muscedere.com.

References

1. R. Muscedere, Improving 2D-Log-Number-System Representations by Use of an Optimal Base, *Eurasip Journal on Advances in Signal Processing*, 2008, 1–13, 2008.
2. V.S. Dimitrov, G.A. Jullien, W.C. Miller, Theory and Applications of the Double-Base Number System, *IEEE Transactions on Computers*, 48, 1098–1106, 1999.
3. R. Muscedere, V. Dimitrov, G. Jullien, W. Miller, A Low-Power Two-Digit Multi-Dimensional Logarithmic Number System Filterbank Architecture for a Digital Hearing Aid, *Eurasip Journal on Applied Signal Processing*, 2005, 3015–3025, 2005.
4. H. Li, G.A. Jullien, V.S. Dimitrov, M. Ahmadi, and W. Miller, A 2-Digit Multidimensional Logarithmic Number System Filterbank for a Digital Hearing Aid Architecture, in *IEEE International Symposium on Circuits and Systems*, ISCAS 2002, 2002, vol. 2, pp. II-760–II-763.
5. V. Dimitrov, S. Sadeghi-Emamchaie, G.A. Jullien, W.C. Miller, A Near Canonic Double-Based Number System (DBNS) with Applications in Digital Signal Processing, in *Proceedings of the SPIE—The International Society for Optical Engineering*, Denver, CO, 1996, pp. 14–25.
6. G.A. Jullien, V.S. Dimitrov, B. Li, W.C. Miller, A. Lee, M. Ahmadi, A Hybrid DBNS Processor for DSP Computation, in *Proceedings of the 1999 IEEE International Symposium on Circuits and Systems*, ISCAS '99, 1999, vol. 1, pp. 5–8.
7. V.S. Dimitrov, J. Eskritt, L. Imbert, G.A. Jullien, W.C. Miller, The Use of the Multi-Dimensional Logarithmic Number System in DSP Applications, in *Proceedings of 15th IEEE Symposium on Computer Arithmetic*, 2001, pp. 247–254.
8. D.M. Lewis, An Architecture for Addition and Subtraction of Long Word Length Numbers in the Logarithmic Number System, *Transactions on Computers*, 39, 1325–1336, 1990.

9

Integrated Circuit Implementations and RALUT Circuit Optimizations

9.1 Introduction

This chapter discusses the chronological evolution of six double-base number system (DBNS)/multidimensional logarithmic number system (MDLNS) hardware designs from the VLSI Research Group at the University of Windsor and the ATIPS Laboratories at the University of Calgary. We also include a final section on circuit level optimizations that evolved as our design techniques matured, and some of this work is based on a 2008 IEEE publication by R. Muscedere [1].

Our early designs used high-level tools to produce rather inefficient layouts, but with the ability to compare designs using more standard binary architectures with the same tools. These early implementations were inefficient in terms of circuit area and performance, but as research into the underlying number system continued, the designs were incrementally improved by incorporating new research features. The designs are all based on a combination of the building blocks introduced in Chapters 5–8.

9.2 A 15th-Order Single-Digit Hybrid DBNS Finite Impulse Response (FIR) Filter

Our first designs, in the late 1990s, were completed using high-level tools, targeting a 0.35 μm complementary metal oxide semiconductor (CMOS) fabrication opportunity offered to Canadian universities at that time. Although a considerable amount of no-cost area was available at that time, there was a rather short submission deadline that limited the design time. A team quickly developed a design that met the early conceptual specifications. This design was by no means optimal; we had limited knowledge of the capabilities of the DBNS index calculus at that time, and the requirement for a fast design turnaround required automated software design approaches. It is important to note that none of the optimization techniques, outlined in Chapter 8, were developed at this point. The process technology and design tools were new to the developers, so this design was also an exercise in developing a submicron design methodology and flow with a new tool set. The design was published in [2] and is briefly discussed here.

9.2.1 Architecture

Digital filter architectures require many additions and multiplications. Each multiplier block represents a filter coefficient, contributing to the order of the filter; thus 15 multiplication blocks represent a 15th-order filter. In order to arrange the multipliers and adders into a parallel architecture, where there are a minimum number of binary two operand adders, a systolic array architecture was used (see Chapter 5 for a complete description). Systolic arrays are a class of pipelined array architectures that feature the properties of modularity, local connectivity, regularity, high level of pipelining, and highly synchronized multiprocessing. Systolic arrays are useful architectures for systems that have multiple operations performed on each data element in a repetitive manner. This gives the system a highly parallel processing architecture. These repeated operational blocks in the systolic array are called processing elements. The choice of a 15th-order system was based on the fabrication area available and initial estimates of the individual processing elements.

9.2.2 Specifications

This first filter design used a hybrid architecture; that is, the input data and filter coefficients used different numbers of digits for their 2DLNS representations. The input 10-bit data were converted using a two-digit 2DLNS representation: 5 bits for the binary exponent, $a_i \in [-16, 15]$, and 4 bits for the ternary exponent, $b_i \in [-8, 7]$. The filter coefficients used a single-digit 2DLNS representation with 10 bits for each of the binary and ternary exponents. The exponent ranges ($a_i \in [-496, 496]$, $b_i \in [-504, 504]$) on the coefficients were limited to avoid DBNS overflow during multiplication. The hybrid design was based on the systolic array architecture discussed in Chapter 5. A 10-bit address LUT was used to convert the input data to a 1-digit 2DLNS, and was connected to 15 pipelined dual-channel computational units (CUs) in series, each channel computing the inner product of the filter coefficients with one of the digits of the data and producing a converted binary output. A binary summer was used to combine the two independent channels (see Figure 9.1). The coefficients were run-time loadable and could be changed at any time. The clock was manually routed in the opposite direction of data flow to eliminate potential clock skew problems.

The design is very large (9 × 16 mm), which is primarily due to the fact that a ROM compiler was not available for this technology. The binary-to-DBNS and the DBNS-to-binary conversion ROMs found in each processor

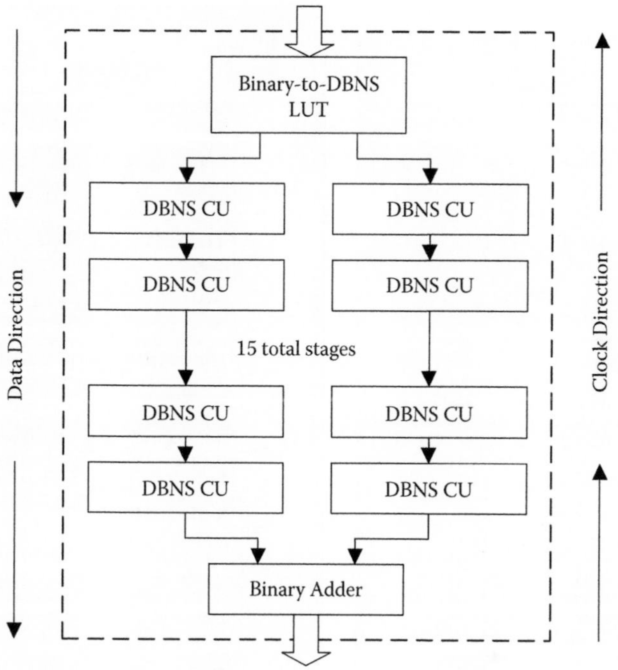

FIGURE 9.1
Architecture of 15th-order single-digit hybrid DBNS FIR filter.

step were implemented using logic gates. Since only three metal layers were accessible in this process, wire routing became quickly congested and the overall cell utilization was very poor, resulting in a very large circuit area.

9.2.3 Results

The layout of the design is shown in Figure 9.2, along with a photomicrograph of the chip. The fabrication turnaround time was approximately 25 weeks (typical for a university), and the chip was tested and found to operate successfully at the designed throughput rate of 20 MHz (the maximum speed of the testing equipment).

9.3 A 53rd-Order Two-Digit DBNS FIR Filter

During the fabrication phase of the previous design, the development team analyzed possible improvements in the design implementation. Clearly the range of the ternary exponents for the coefficients was very wide, resulting in very large DBNS-to-binary tables. To reduce this range a full two-digit

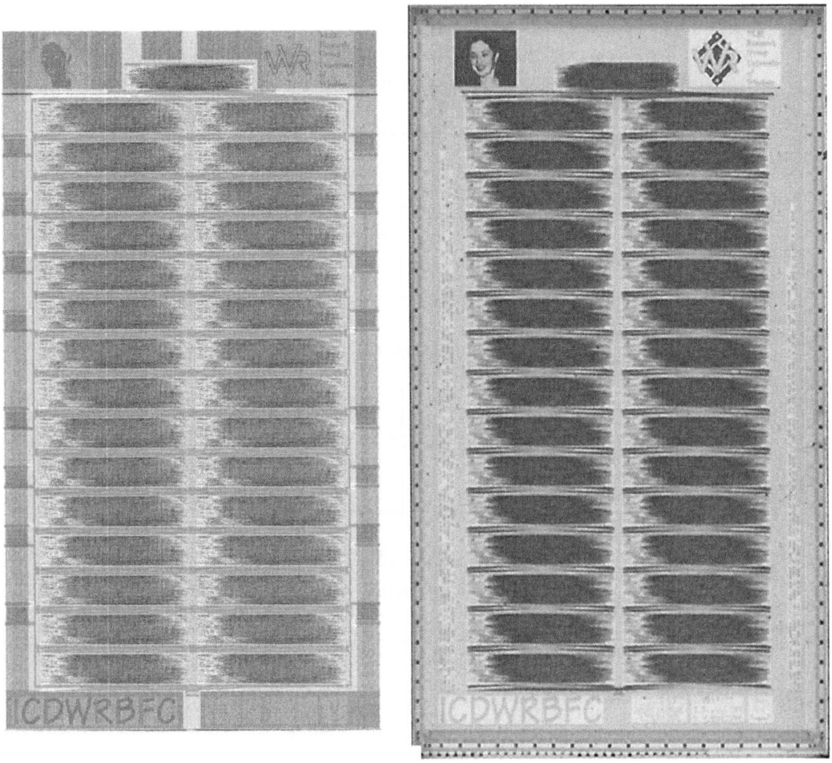

FIGURE 9.2 (See color insert)
Fifteenth-order single-digit hybrid DBNS FIR filter chip (layout and photomicrograph).

architecture was proposed. The resulting product of the two digits would generate a four-digit DBNS result; however, given that both the data and coefficients were encoded using the "greedy" algorithm, the fourth channel, which is the least significant, was removed (see Section 5.5.4 in Chapter 5).

The architecture of this two-digit 2DLNS filter used the same systolic array structure as the hybrid filter design. Using the same I/O specifications as for the hybrid filter, an I/O bound 53rd-order filter could be implemented.

9.3.1 Specifications for the Two-Digit 2DLNS Filter

This filter used a two-digit DBNS representation for both the 10-bit input data and the filter coefficients. As in the hybrid design the two-digit 2DLNS representation used 5 bits for the binary exponents and 4 bits for the ternary exponents. The exponent ranges on the coefficients were again limited to avoid DBNS overflow during multiplication. The resulting DBNS-to-binary tables were considerably smaller; however, the limitation on the number of routing layers still limited the possible reduction in silicon area.

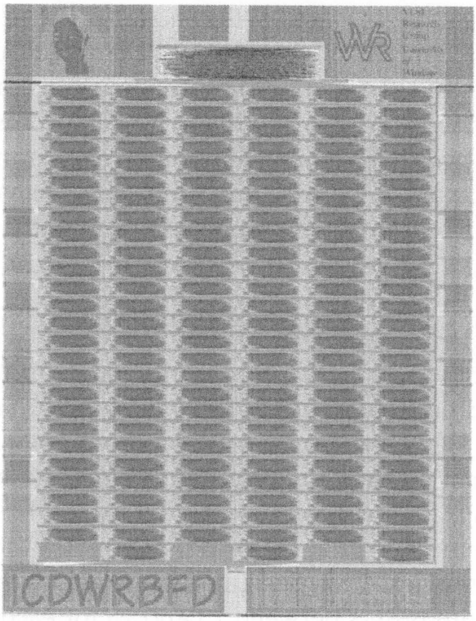

FIGURE 9.3 (See color insert)
Fifty-third-order two-digit DBNS FIR filter (layout only).

9.3.2 Results

A layout of the design (which was 9 × 11.4 mm) is shown in Figure 9.3. Based on the excellent agreement of the simulation versus fabrication results of the hybrid design, the two-digit design was analyzed at the simulation level only, where the simulations showed that the design worked as expected [3]. The computational unit (CU) was 9.5 times smaller than the previous hybrid design, resulting in many more CUs able to be placed on a single chip, thus allowing for a considerable increase in the filter order.

9.4 A 73rd-Order Low-Power Two-Digit MDLNS Eight-Channel Filterbank

From the results of the previous two filter designs, a significant reduction in hardware area in the DBNS-to-binary converter occurs when moving from a hybrid architecture to a full two-digit 2DLNS design; however, the overall area is still quite large. In order to further reduce the size of the filter architecture, a serial computational approach was considered and explored in [4] with the intention of reducing power consumption by reducing the processing speed. Rather than processing one output per clock cycle, as in the previous designs, this design performs a single multiply-accumulate (MAC) per clock. The design uses a state machine to control buffering of the filter inputs, routing of the inputs and outputs of the MAC, and outputting the final filter results. Just as in the previous design, the fourth channel was removed. Unlike the two previous designs, a 0.18 μm CMOS process technology was used that provided five layers of metal, therefore alleviating the routing congestion. Lastly, a binary-to-DBNS conversion circuit was developed at the same time in order to eliminate the large LUT used in the first two designs.

9.4.1 Specifications

This filter used a two-digit DBNS representation for both the 16-bit input data and the filter coefficients: 5-bit binary and 4-bit ternary exponents ($a_i \in [-12,12]$, $b_i \in [-4,3]$). These specifications use a shorter binary word to encode the DBNS representation than the original designs, but at the cost of accuracy. However, this is justified as the intended use of the filter is in hearing instruments where the dynamic range is more important than the absolute error of the representation. This design implemented eight independent filters (a filterbank) with orders between 73 and 75 for each to meet the 60 dB stop band requirement. The design requires 625 clock cycles to process eight outputs, for a 16 kHz sampling frequency.

9.4.2 Results

Simulation results showed that this design consumed approximately 1.175 mW at 10 MHz (117 μW/MHz) with a 1.6 V supply. This design was competitive at the time with other published work using different architectures and fabrication processes. The chip was not fabricated since significant advancements were continually being made during its development.

9.4.3 Improvements

Several improvements were made during the development of the initial design. They are briefly listed here, but for a full discussion see [5] and [6].

- Set all the filter orders to 75 to allow for using evenly spaced pass bandwidths for each filter. This results in only having to store 152 symmetrical FIR coefficients as opposed to the 592 coefficients of the original filterbank.
- Symmetrical filters can be processed in parallel using dual CUs, therefore halving the operating frequency (625 cycles to 313 cycles).
- Use the optimal base algorithm (as discussed in Chapter 8) to optimize the filter coefficient mapping. This results in a filter coefficient mapping of 6 bits for the binary exponent, and only 2 bits for the ternary exponent, and a mapping of 6 bits for the data binary exponent

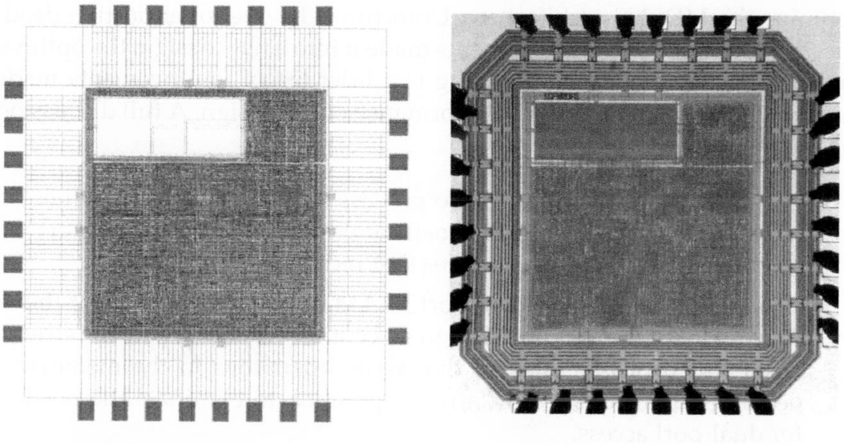

FIGURE 9.4 (See color insert)
Seventy-third-order low-power two-digit MDLNS eight-channel filterbank (layout and photomicrograph). (From H. Li, G.A. Jullien, V.S. Dimitrov, M. Ahmadi, W. Miller, A 2-Digit Multidimensional Logarithmic Number System Filterbank for a Digital Hearing Aid Architecture, in *IEEE International Symposium on Circuits and Systems*, ISCAS 2002, 2002, vol. 2, pp. II-760–II-763. Copyright © 2002 IEEE.)

with a 5-bit mapping for the ternary exponent. The 5-bit ternary exponent is limited to the range $b_i \in [-14,14]$ to avoid overflow.

- Use the full 4-digit MDLNS output to minimize output error.

9.4.4 Results

The layout and photograph of the design are shown in Figure 9.4. It was successfully tested and had a power consumption of 708 μW at 5 MHz (or 141.6 μW/MHz), which was in good agreement with simulation estimates. The power is higher than the initial design since it is performing dual computations per clock. The overall area was 1.67 × 1.67 mm with I/O pads, but only 1 × 1 mm for the core of the design.

9.5 Optimized 75th-Order Low-Power Two-Digit MDLNS Eight-Channel Filterbank

The previous design provided confirmation that the MDLNS (a 2DLNS in this case) could be used for filterbank applications and possibly save power in the process. Although the design was essentially a collection of existing MDLNS building blocks, the power results were encouraging enough for us to attempt a new design. During the time the original design was being implemented, parallel research work was being done to further improve and optimize the MDLNS architecture. Unfortunately, due to fabrication deadlines, only one of these improvements made it into a new design: the optimal base selection system. The following is a brief list of improvements made that substantially increased the performance of the design. A full discussion can be found in [7].

- The controller was improved to process any even number of filterbanks and any odd number of coefficients through the use of "smart" counters and dynamic references that reduced the overall logic.
- The design utilized a single-port SRAM exclusively in favor of the previously used dual-port design since it is essentially half the size. A pipelined design ensures that write operations to the memory occur simultaneously with a read operation to eliminate the need for dual-port access.
- Dual-phase clocking to synchronous SRAM was removed, thereby reducing the power by eliminating extraneous circuit switching.
- All maintenance/timing cycles were removed from the controller to obtain a true 300-cycle system, thereby requiring only a 4.8 MHz clock.
- The CU was streamlined as well as the four-channel accumulator by using the single-bit sign architecture (introduced in Chapter 8). This also appreciably reduced power by eliminating unnecessary switching.
- The symmetrical filter output was generated by adding or subtracting the four-channel output based on the symmetry of the coefficients.
- The second base was optimized by giving a higher priority to the input data mapping than that of the filter coefficients. This resulted in a slight increase in the filter coefficient mapping range by increasing the ternary exponent range to 3 bits ($b_i \in [-3, 3]$) and decreased mapping for the data coefficients' ternary exponent range ($b_i \in [-12, 12]$). Although the data mapping is reduced, the representation is considerably improved. Originally, 18,348 of the 32,768 possible values were error-free; the remainder had representation errors between 0.5 and 37. By optimizing the data, 19,513 of the 32,768 possible values are error-free with the worst-case error below 6.

9.5.1 Results

Given that the design flow and methodology were proven to be accurate for power estimations from past designs, this design was tested by simulation only, which demonstrated that this design was superior in many aspects. The power was shown to be 316 μW at 4.8 MHz (65 μW/MHz) for a worst-case scenario, a savings of over 55%. Assuming normal operation, the power dropped as low as 180 μW (37.5 μW/MHz). The overall logic cells were reduced in area by 47% with an estimated core area of approximately 600 × 600 μm.

9.6 A RISC-Based CPU with 2DLNS Signal Processing Extensions

In order to ease the implementation of MDLNS in DSP applications, and to minimize design and simulation times, the concept of having a complete CPU based on MDLNS was considered. The designed CPU implements arithmetic operations such as multiply and accumulation (MAC) as a special instruction in its instruction set running on a MDLNS platform. Every multiplication in this CPU is performed on MDLNS data, which are then converted back to binary for addition and subtraction. The CPU can be considered to be relatively simple reduced instruction set computer (RISC) architecture, which performs MDLNS/binary conversions, MDLNS multiplication and MAC, and other tasks in addition to most traditional arithmetic and logical operations. The reasonably small number of instructions, limited instruction types, and simple instruction architecture provides the CPU with a simple assembly language, and makes it applicable to other research work, as well as the realization of DSP algorithms. Essentially every application is an assembly program running on this CPU. The arithmetic and logic unit (ALU) performs most of the traditional arithmetic and logic operations, while the other components can be used to perform special instructions based on 2DLNS operations. Taking advantage of a fundamental MAC unit, the CPU can be programmed to realize vector inner products, convolution, FFT, and filtering processes that heavily rely on optimized MAC operations. The design is fully detailed in [8] and [9].

9.6.1 Architecture and Specifications

The 2DLNS CPU is composed of registers, buses, multiplexers, an ALU, a MAC unit, a binary/2DLNS converter (BTC), and a sequential control unit. Figure 9.5 shows the architecture of the CPU.

The controller sequences the data path operations by performing a set of fetch, decode, and execute stages for every CPU instruction. Based on the instructions provided, the controller moves data between each of the functional blocks. The CPU has 16 general purpose registers and 64 total instructions.

The MDLNS portion of the CPU uses the same optimal base and exponent ranges of the optimized filterbank design. For this two-digit MDLNS, this results in a 24-bit word that is mapped into the widths of the two memories (instruction and data).

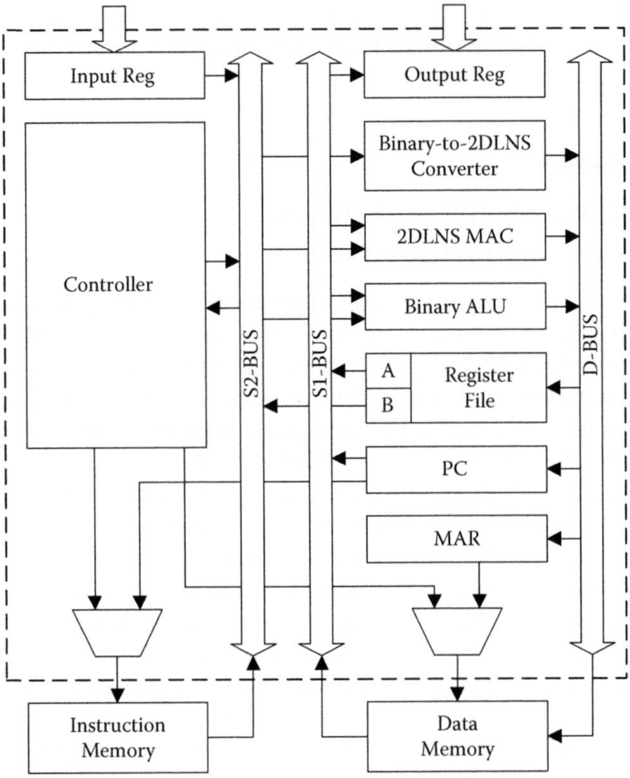

FIGURE 9.5
Architecture of the MDLNS CPU.

9.6.2 Results

As a test of the CPU design, the same filterbank from the optimized design was implemented; in total, only 55 instruction words were used. What is more important is that this code was written very quickly (about one day), whereas the optimize filterbank took over a month to design and verify. The comparison is shown in Table 9.1 excluding the third-party SRAM block as the power could not be measured. As expected, the CPU version requires more power as well as area since it includes more registers and performs more switching to achieve the same results. The filterbank operation in the 2DLNS CPU requires 328 clock cycles; the remaining 266 are required for processing overhead (input conversions, loops, control, and output). However, the objective of this design is not simply to implement a filterbank, but any potential application. Implementing such applications on this CPU requires no HDL programming experience, with only a working knowledge of the CPU instruction set being sufficient to achieve results in a considerably shorter time than a full custom approach. This CPU-based design can be further improved to be more competitive in regards to the timing and

TABLE 9.1

Comparison of Filterbank Implementations

Design	Improved Filterbank [4]	Optimized Filterbank [6]	MDLNS CPU [7]
Interconnects	5,759	4,877	7,176
Cells	7,005	3,742	7,056
Cell area (μm²)	184,965	53,716	206,812
Power (μW)	708 @ 5 MHz	316 @ 4.8 MHz	990 @ 9.5 MHz
Normalized power (μW/MHz)	141.6	65	104.2

power consumption by adding more specialized instructions, but not to the degree that it is transformed into an application-specific processor.

9.7 A Dynamic Address Decode Circuit for Implementing Range Addressable Look-Up Tables

The range addressable lookup table (RALUT) is a nonlinear memory storage element shown in Chapter 6 to significantly reduce hardware requirements for matching data in short and very long words. Unfortunately most of the circuits utilizing RALUTs, presented thus far, are built with logic gates and tristate buffers so that they are easily synthesizable and implemented with other components of the overall design. These circuits will be more viable if the RALUT implementation were competitive with modern memory in terms of area, timing, power, and functionality. Recall that the difference between a RALUT and a standard LUT is the address decoding system; therefore only this portion of the design will be analyzed here. There are certainly many other components in modern memory design [10] (e.g., self-timing circuits, redundancy, self-testing) that will not be considered here. The actual storage systems of the output data (read only, read/write) and mechanisms (static, dynamic) will be the same between LUTs and RALUTs. The RALUT implementations we present will be read only; however, it is possible to make them fully programmable. This section shows a preliminary dynamic address decode circuit that can be used to build a scalable full custom RALUT.

9.7.1 Efficient RALUT Implementation

Both discrete and ASIC memory components are customized and tuned for the technology in which they are implemented to achieve the highest performance (whether it is speed, size, reliability, or power). Memory fabrication is specialized where single foundries are designed and tuned for a particular type and size of memory. Because of the proprietary nature of the memory industry, it is difficult to find any published literature that discusses all details of state-of-the-art memory design. The following section will detail one method for implementing a RALUT address decode using variations of published common circuits [11,12].

Although it is possible to implement large LUTs with only logic gates, it is certainly not efficient since the resulting circuit will have an excessive amount of logic cells and interconnects. The physical implementation will most likely be inefficient as the place and route tools are designed to optimally place groups of cells via hierarchical designs. Most synthesis tools should only use logic gates for very small LUTs as their implementation cost is less than LUT IP blocks, which may contain extra overhead such as additional clock cycles. The same can be said for a RALUT. In [4], more than a quarter of the chip core area is devoted to the logic gate version of the RALUT.

9.7.2 Dynamic Address Decoders (LUT)

Memory circuits use dynamic address decoders to efficiently determine the proper word line to enable for memory access. One of the most common methods of decoding is to first use predecoder circuits to shorten the main word line logic [11]. As an example, a 4-bit LUT address decode circuit is shown. The high (see Figure 9.6) and low (see Figure 9.7) address words are first predecoded from 2 bits (binary) to 4 bits (one hot encoding; only 1 of 4 bits is high at one time). The input RA3[0:1] is memory circuit terminology that refers to all states of address bits 3 (A3 & $\overline{\text{A3}}$) and is denoted by [0:1] or 0 to 1. Since the high predecode uses both RA3[0:1] and RA2[0:1], there are four possibilities resulting in four predecoded lines RA32L[0:3]. The chain of NMOS transistors is not usually long (this usually depends on the technology), as this can cause problems related to charge leakage and sharing as well as cross talk issues [13].

The address decoders are activated by the driving clock, which is usually derived from the system clock. Depending on the density of the memories,

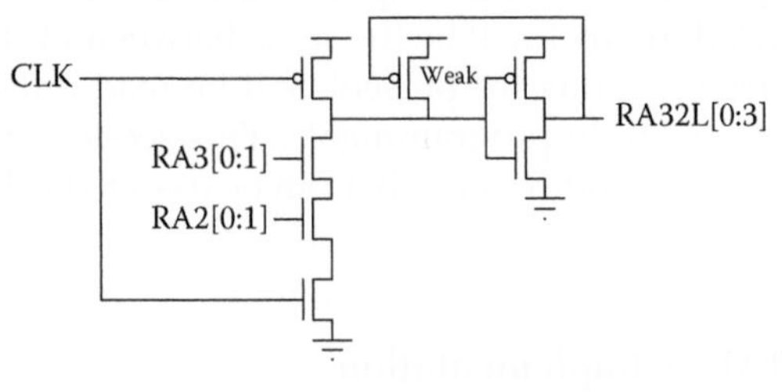

FIGURE 9.6
Four predecoders (high word) for LUT. (From R. Muscedere, K. Leboeuf, A Dynamic Address Decode Circuit for Implementing Range Addressable Look-Up Tables, in *IEEE International Symposium on Circuits and Systems*, ISCAS 2008, 2008, pp. 3326–3329. Copyright © 2008 IEEE.)

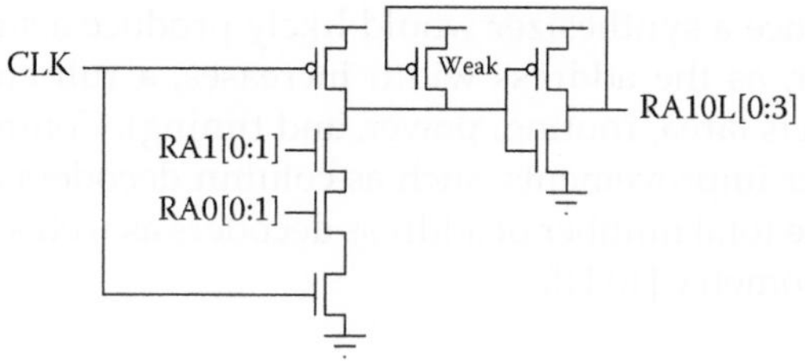

FIGURE 9.7
Four predecoders (low word) for LUT. (From R. Muscedere, K. Leboeuf, A Dynamic Address Decode Circuit for Implementing Range Addressable Look-Up Tables, in *IEEE International Symposium on Circuits and Systems*, ISCAS 2008, 2008, pp. 3326–3329. Copyright © 2008 IEEE.)

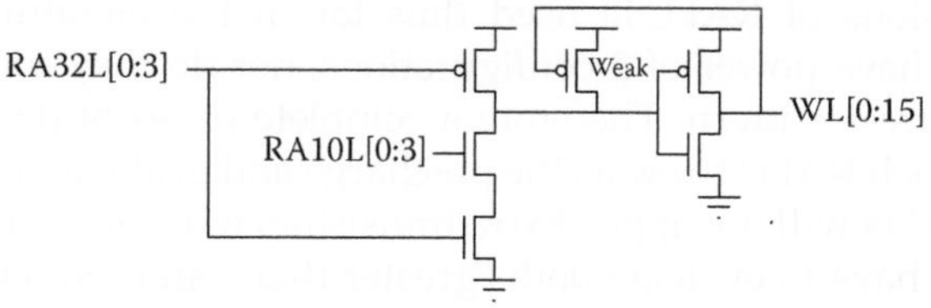

FIGURE 9.8
Sixteen second-stage decoders for LUT. (From R. Muscedere, K. Leboeuf, A Dynamic Address Decode Circuit for Implementing Range Addressable Look-Up Tables, in *IEEE International Symposium on Circuits and Systems*, ISCAS 2008, 2008, pp. 3326–3329. Copyright © 2008 IEEE.)

it is possible that the driving clock could be two to four times faster than the system clock. When the driving clock is low, the system is in a precharge state where the output of all the decode trees is low. During this time, the address bits are set, which allows some time for the channels of the NMOS transistors to stabilize prior to the evaluation state. The output of the decoder is fed back to a bit keeper to maintain high logic levels and slightly faster signal pull-up for precharge. This circuit is duplicated for the low-word address also (address bits 1 and 0).

At this point two groups of four lines contain the one hot encoding of the high and low address words. These lines are fed into a second-stage decoder to generate the final word lines (see Figure 9.8). Depending on the size of the address space, there may be more stages of decoding (i.e., third, fourth, etc.). In the second-stage decoder, the driving clock used in the first stage is ignored. Instead, one of the bits from the high predecoder acts as the clock driving the circuit. If the high predecode output for that particular bit is low, then the corresponding word line would not be active and no transistor switching occurs, in essence, automatic power savings. Four of the 16 second-stage decoders will actually enter dynamic evaluation mode, of which only one will output high. This lone word line will activate the output logic, which is usually configured safely as a "wired or" network.

For this example, a 4-bit dynamic address decoder, only 72 PMOS and 80 NMOS transistors (excluding any buffers) are required. This, of course, is not optimal since a synthesizer would likely produce a smaller gate level design. However, as the address width increases, a full custom approach saves in all aspects (area, routing, power, and timing). Commercial memory design uses other improvements, such as column decoders and bank selectors, to reduce the total number of address decoders as well as to improve the overall layout geometry [10,11].

9.7.3 Dynamic Range Address Decoders (RALUT)

Using some of the above concepts of dynamic address decoders, preliminary dynamic range address decoders can be implemented so that it is possible to build a RALUT on par with dynamic LUT circuits.

The dimensions of RALUTs used thus far in the literature have been shown not to have power of 2 configurations, nor do they conform to any particular address pattern. Therefore a complete series of decodes will be required for each RALUT row as the precharge and evaluate tree reductions in standard LUTs will not apply. Extra transistors will also be needed as the decoders will have to evaluate both "greater than" and "equal" conditions. Figure 9.9 shows the RALUT decode logic with special attention to the transistor configurations for comparing address lines to 0's and 1's. When A[n] is compared with 1, only a single transistor is used, similar to a LUT address decode. However, when A[n] is compared to 0, two transistors are needed as one branches to the "greater than" line.

As mentioned previously, a large chain of NMOS transistors can cause charge leakage and sharing problems. One of the benefits of the RALUT is its ability to match very large words to very few outputs. From Figure 9.9 the number of NMOS transistors in these chains will be the same as the number of bits in the address (excluding the clock). For sizable input buses, these address chains will need to be broken up. Unlike the standard LUT, the

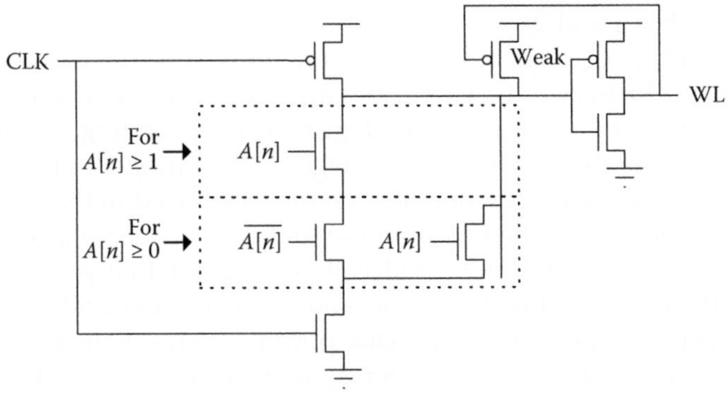

FIGURE 9.9
RALUT decode logic. (From R. Muscedere, K. Leboeuf, A Dynamic Address Decode Circuit for Implementing Range Addressable Look-Up Tables, in *IEEE International Symposium on Circuits and Systems*, ISCAS 2008, 2008, pp. 3326–3329. Copyright © 2008 IEEE.)

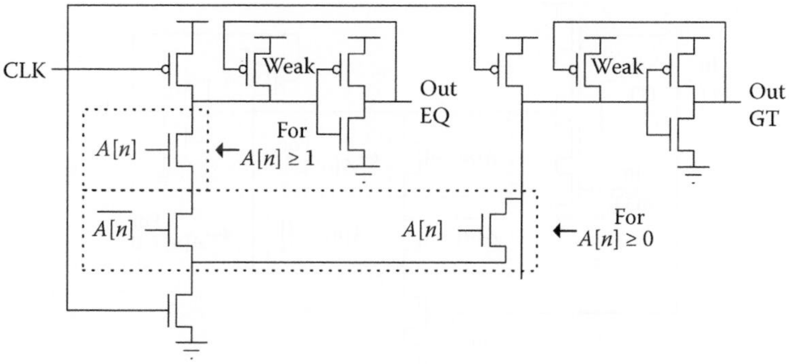

FIGURE 9.10
RALUT first-stage decode logic. (From R. Muscedere, K. Leboeuf, A Dynamic Address Decode Circuit for Implementing Range Addressable Look-Up Tables, in *IEEE International Symposium on Circuits and Systems*, ISCAS 2008, 2008, pp. 3326–3329. Copyright © 2008 IEEE.)

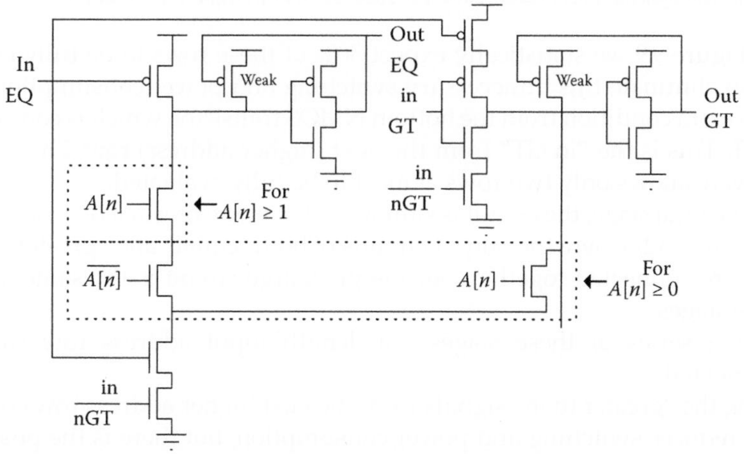

FIGURE 9.11
RALUT middle-stage decode logic. (From R. Muscedere, K. Leboeuf, A Dynamic Address Decode Circuit for Implementing Range Addressable Look-Up Tables, in *IEEE International Symposium on Circuits and Systems*, ISCAS 2008, 2008, pp. 3326–3329. Copyright © 2008 IEEE.)

decode stages in a RALUT cannot be easily interconnected as the "greater than" and "equal" signals must remain separated until the last stage. Extra transistors must be added to properly pass information from one stage to another. Figure 9.10 shows the first stage of a multistage implementation. Here the "greater than" and "equal" precharge and output circuits are separated.

For the middle stages (see Figure 9.11) the "equal" precharge and output circuit remains the same. The "greater than" precharge evaluates high immediately if the previous "greater than" stage was also high, whereas if the previous "equal" stage was low, the "greater than" stage is never precharged.

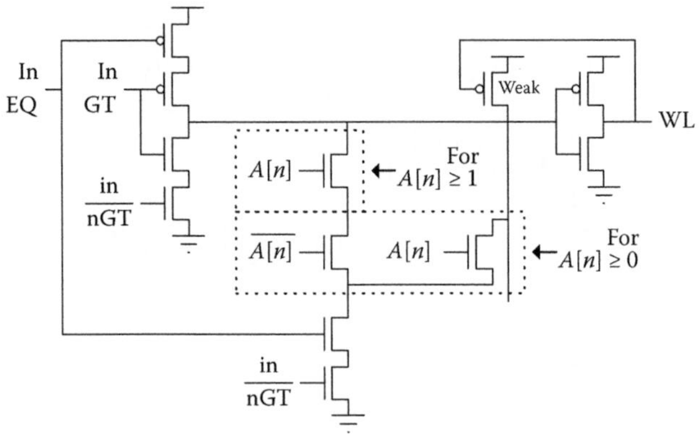

FIGURE 9.12
RALUT final-stage decode logic. (From R. Muscedere, K. Leboeuf, A Dynamic Address Decode Circuit for Implementing Range Addressable Look-Up Tables, in *IEEE International Symposium on Circuits and Systems*, ISCAS 2008, 2008, pp. 3326–3329. Copyright © 2008 IEEE.)

From Figure 9.9 we statistically expect 50% of these rows to be fully evaluated. To eliminate this unnecessary switching and power consumption, we add an extra condition from the bottom NMOS transistor, which is controlled by nGT. This is the "in GT" from the next (higher address) row. This signal effectively allows only two rows at most to be fully evaluated.

For the final stage, the circuit is similar to that of the original decode circuit (see Figure 9.12) where the output circuits of both "equal" and "greater than" stages are connected together, but the precharge circuit is the same as the middle stages.

With a series of these stages any length input address row can be implemented.

Using the "greater than" signals from the next higher address row considerably reduces switching and power consumption, but there is the possibility of two adjacent word lines being active at the same time. This condition is eliminated by using a simple logic gate to ensure that only one word line is active. This word line is connected to the "wired or" safe network, which generates the RALUT outputs.

9.7.4 Full Custom Implementation

To simplify the implementation of RALUTs, an autogeneration tool has been developed in SKILL that operates in the Cadence Design Framework II environment. It uses several predesigned parameterized cells so that the resulting RALUT layout and transistor parameters can be easily adjusted as well as scaled for smaller process technologies. Other parameters such as the number of input bits per stage, and clock/input buffer tree frequency are also adjustable. Figure 9.13 shows the layout of this RALUT implementation.

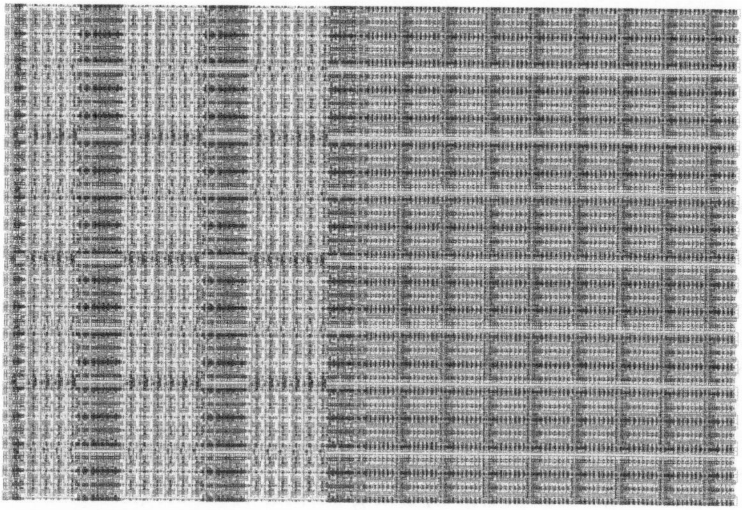

FIGURE 9.13 (See color insert)
Screen dump of a 29 × 16 × 52 RALUT (0.18 μm TSMC [Taiwan Semiconductor Manufacturing Company] process). (From R. Muscedere, K. Leboeuf, A Dynamic Address Decode Circuit for Implementing Range Addressable Look-Up Tables, in *IEEE International Symposium on Circuits and Systems*, ISCAS 2008, 2008, pp. 3326–3329. Copyright © 2008 IEEE.)

TABLE 9.2
Comparison to Published Synthesized Design

	Unrouted Area (μ m^2)	Routed Area (μ m^2)	Critical Path (ns)	Power Consumption (μW/MHz)
[4]	71,786	257,400	4.45	52
Proposed	41,778	41,778	2.70	25
Savings	41.8%	83.8%	39.3%	52.0%

9.7.5 Comparison with Past Designs

We can compare our proposed full custom layout with a previously published [4] synthesized implementation using Virtual Silicon standard cells. The RALUT is the same geometry (29 rows, 16 input bits, 52 output bits). The results are shown in Table 9.2. The proposed circuit offers considerable improvements in all aspects. The routed area in [4] is large to accommodate for the routing congestion discussed earlier.

9.7.6 Additional Optimizations

For larger RALUT configurations, the circuit can be subdivided into smaller units or banks based on the most significant bit (MSB) of the input address. The MSBs can be decoded using the standard LUT predecoder circuits

(Figures 9.6 to 9.8) to activate only one RALUT bank in particular. This would save some area in terms of the address decoders, and it would also save power since only one bank would evaluate, as opposed to all of them.

9.8 Summary

This chapter concludes the DSP thrust for targeted applications for DBNS and MDLNS. The chapter is included because it chronicles the work by our group, over the past decade, in the silicon implementation of several of the architectures that have been developed in the preceding chapters. University laboratories are not always able to access the most current silicon technologies, and it is important that our designs are studied in comparison with more conventional architectures, using similar technologies. This is often a difficult exercise when studying design techniques that are not in mainstream use. The designs we have included demonstrate the efficiencies that can be obtained by using the various optimization and well-chosen design strategies for the architecture and layout of the processors. A CPU-based design that uses MDLNS to advantage in its instruction set with targeted applications in low-power filter implementation caps this part of the chapter.

In the final section, we have discussed a dynamic address decode circuit for implementing a full custom read-only RALUT memory device. In comparison to a published synthesized result, this implementation is at least 41% smaller in area, 40% faster in speed, and uses 50% less power.

The RALUT generator tool has been developed in SKILL, which operates in the Cadence Design Framework II environment. Several parameters can be easily adjusted to tune the design to achieve the desired performance.

All software, source files, and detailed results can be found at http://research.muscedere.com.

References

1. R. Muscedere, K. Leboeuf, A Dynamic Address Decode Circuit for Implementing Range Addressable Look-Up Tables, in *IEEE International Symposium on Circuits and Systems*, ISCAS 2008, 2008, pp. 3326–3329.
2. J. Eskritt, R. Muscedere, G.A. Jullien, V.S. Dimitrov, W.C. Miller, A 2-Digit DBNS Filter Architecture, in *IEEE Workshop on Signal Processing Systems*, SiPS 2000, 2000, pp. 447–456.
3. J. Eskritt, Inner Product Computational Architectures Using the Double Base Number System, MASc, University of Windsor, 2001.

4. H. Li, G.A. Jullien, V.S. Dimitrov, M. Ahmadi, W. Miller, A 2-Digit Multidimensional Logarithmic Number System Filterbank for a Digital Hearing Aid Architecture, in *IEEE International Symposium on Circuits and Systems*, ISCAS 2002, 2002, vol. 2, pp. II-760–II-763.
5. H. Li, R. Muscedere, G.A. Jullien, V.S. Dimitrv, The Application of 2-D Logarithms to Low-Power Hearing-Aid Processors, in *45th Midwest Symposium on Circuits and Systems*, MWSCAS-2002, 2002, vol. 3, pp. III-13–III-16.
6. H. Li, A 2-Digit Multi-Dimensional Logarithmic Number System Filterbank Processor for a Digital Hearing Aid, MASc, University of Windsor, 2003.
7. R. Muscedere, V. Dimitrov, G. Jullien, W. Miller, A Low-Power Two-Digit Multi-Dimensional Logarithmic Number System Filterbank Architecture for a Digital Hearing Aid, *Eurasip Journal on Applied Signal Processing*, 2005, 3015–3025, 2005.
8. M. Azarmehr, R. Muscedere, A RISC Architecture for 2DLNS-Based Signal Processing, *International Journal of High Performance Systems Architecture*, 3, 149–156, 2011.
9. M. Azarmehr, A Multi-Dimensional Logarithmic Number System Based Central Processing Unit, University of Windsor, 2007.
10. K. Itoh, K. Osada, T. Kawahara, Trends in Low-Voltage Embedded RAMs, in *2nd Annual IEEE Northeast Workshop on Circuits and Systems*, NEWCAS 2004, 2004, pp. 45–48.
11. J.S. Caravella, A Low Voltage SRAM for Embedded Applications, *IEEE Journal of Solid-State Circuits*, 32, 428–432, 1997.
12. G. Samson, N. Ananthapadmanabhan, S.A. Badrudduza, L.T. Clark, Low-Power Dynamic Memory Word Line Decoding for Static Random Access Memories, *IEEE Journal of Solid-State Circuits*, 43, 2524–2532, 2008.
13. P. Srivastava, A. Pua, L. Welch, Issues in the Design of Domino Logic Circuits, in *Proceedings of the 8th Great Lakes Symposium*, VLSI 1998, 1998, pp. 108–112.

10

Exponentiation Using Binary-Fermat Number Representations

10.1 Introduction

In this final chapter, we discuss the use of the double-base number system (DBNS) in performing exponentiation over finite fields, in particular, Galois fields, since the importance of fast exponentiation over such fields for modern cryptography is very high. The most famous and widely used cryptosystem, RSA [1], and many other number theoretic systems [2–7] rely on the existence of fast methods for exponentiation, and a large variety of techniques can be found in the literature. Although deeply studied, this computational operation still possesses some unknown features, the solutions of which might have a significant impact on the efficient performance of many cryptosystems.

10.1.1 Some Examples of Exponentiation Techniques

- It is generally accepted [20,24,30] that the parallel complexity of modular exponentiation is not well understood, and yet efficient parallel algorithms are crucially relevant to VLSI implementation, for example, in smart cards.
- Several commercial cryptosystems [8,20] make use of an exponentiation over $GF(2^n)$. In these cases squaring can be assumed to be an almost free operation, if one uses the so-called normal bases representation, because squaring reduces to cyclic bit shifts. Therefore we can look for methods that reduce, as much as possible, the number of regular multiplications at the price of an almost unlimited increase in the number of squarings. Agnew et al. [8] have used this consideration in their algorithm for performing arithmetic over $GF(2^n)$, and Newbridge Corp. has used this algorithm in its cryptochip operating over $GF(2^{593})$.
- The role of precomputations has been mentioned several times in the literature, but it is still only mainly used in systems such as the Diffie-Hellman key exchange algorithm [2] or the El-Gamal digital signature algorithm [5].

- From a hardware viewpoint, the space (area) complexity of the algorithms is at least as important as the time complexity, especially in the case of severe memory restrictions. Most of the algorithms based on the sliding window approach tend to use too much storage, although this is not often a consideration in the literature promoting such techniques. In the case of smart cards this is definitely an important component of the cost function.
- The general formulation of the problem for evaluation of modular exponentiation is: given x, y, and z, compute $x^y (mod\ z)$. In reality z is usually a constant (we are unaware of any cryptosystem using variable moduli). One of the two remaining parameters is a constant (a public or private key), and the last one constitutes the message to be encrypted. The Diffie-Hellman key exchange algorithm and the DSS and El-Gamal cryptosystems use a fixed base and a variable exponent. The RSA algorithm uses a fixed exponent and variable base. The speed of fixed-base cryptosystems can be increased by precomputing and storing certain powers of the base. This also includes negative powers, that is, computing certain inverse elements. In the case of RSA this consideration does not work, but one can spend some efforts in attempt to find a "nice" exponent representation that will reduce the amount of time needed to perform exponentiation.
- The invention of elliptic curve cryptosystems (ECCs) [3] has led to significant changes in cryptography. Practically every number theoretic cryptosystem can be translated into the language of elliptic curves over finite fields/rings. In ECC one uses multiplication instead of exponentiation. Therefore the main implementation difference between fixed-base and variable-base cryptosystems, defined over finite fields/rings, becomes irrelevant in this case. From an algorithmic viewpoint, this leads to two important conclusions: first, one can get substantial improvements by finding a good representation of the multiplier known in advance; and second, the operation to negate a point is cheap, and one can use addition-subtraction chains instead of simply addition chains. This extra flexibility permits shorter chains.

10.1.2 About This Chapter

A better understanding of all these issues will lead to faster and more reliable methods for exponentiation. At the present time, it is particularly important considering the recently increased security measures that require the use of extremely large exponents (e.g., 768 bits for Visa credit cards).

These considerations are the driving force of this chapter. To summarize the results of our efforts we will demonstrate a recently developed method that leads to algorithms that use very few regular multiplications.

It is particularly useful for every cryptosystem based on GF(2^n) arithmetic. Compared to the known algorithms, the technique proposed is about 10% faster than the best reported algorithms. What seems to us much more important, however, is that our approach uses very few (three) registers. Our novel algorithm is based on certain new properties of the double-base number system [16], and from a purely number theoretic viewpoint they provide a new insight to this number representation scheme.

10.2 Theoretical Background

The problem in minimizing the number of modular multiplications for exponentiation is closely related to the so-called addition chains. An addition chain [26] for a given integer, t, is a succession of positive integers: $a_1 = 1$, $a_2, \ldots, a_l = t$ such that for every $p > 1$: $a_p = a_j + a_k$ for some k and j, $1 < j \leq k < l$.

Obtaining at least one of the shortest addition chains for a given integer is an NP-complete problem [13,29]. The lower bound for the shortest addition chain length was established by Shönhage, in Knuth [12]. His result states that no addition chain for t can be shorter than $\log_2 t + \log_2 H(t) - 2.13$, where $H(t)$ is the Hamming weight of t.

This interpretation of the problem has one serious shortcoming—it does not distinguish between multiplications and squarings. This is absolutely crucial if one deals with exponentiations over GF(2^n). Since the elliptic curve cryptosystems over GF(2^n) are gaining in popularity [9–11,25,27,28,31], addressing this issue becomes quite relevant. Some even-characteristic Galois fields of practical importance in cryptography are GF(2^{155}) [32,33], GF(2^{163}) [34] (an encryption algorithm based on ECC over GF(2^{163}) has been proposed for the upcoming IEEE P1363 standard), GF(2^{176}) [34,35], and GF(2^{593}) [8] (fabricated by Newbridge Microsystems Corporation). The examples given in this chapter are based on computational experiments with 593-bit integers; that is, they can straightforwardly be applied to the latter case. Generally speaking, for our approach, the larger the field, the greater the computational savings.

It is clear that if the exponent is a power of 2, then we only need squarings to perform exponentiation. If the exponent is a number of the form $2^n + 1$, we need n squarings and one regular multiplication. This very simple observation leads to the study of the following representation:

$$t = \sum_{i=1}^{d} 2^{a(i)} F_k^{b(i)}, \qquad F_k = 2^k + 1 \tag{10.1}$$

In what follows we shall call this a *binary-Fermat representation*. The integers of the form $2^a F_k^b$, $F_k = 2^k + 1$, a, b - nonnegative integers will be called *binary-Fermat numbers*.

Such a representation can be viewed as a generalization of the DBNS. All the initial applications of DBNS in cryptosystems [14–16] were dedicated to fixed-base cryptosystems. The aim now is *exactly the opposite*—we want to show how to improve the performance of fixed-exponent cryptosystems by making use of an appropriate choice of the exponent in the form of Equation (10.1).

We shall restrict ourselves to the case of odd bases having a Hamming weight of 2. However, it is worth investigating other bases with very small Hamming weights.

Suppose we have some representation of the exponent of the form of Equation (10.1). Then the number of regular multiplications corresponding to this particular representation is given by Equation (10.2).

$$RM(t) = \underbrace{\max}_{i=1,2,\ldots,d} b^{(i)} + d - 1 \tag{10.2}$$

In this formula, d is the number of binary-Fermat numbers, such that their sum is exactly t, by using Equation (10.1).

Finding a representation that corresponds to the global minimum of $RM(t)$ is a challenging problem, and appears, at the current time, to be very difficult to solve. Instead, we offer heuristic rules on how to obtain fairly good representations that lead to exponentiations with very few regular multiplications.

One of the simplest ways (but by no means the only way) to compute good representations of the form of Equation (10.1) is to use the greedy algorithm that we have made considerable use of in earlier chapters. Providing that the odd (Fermat) base is fixed, the algorithm is structured as shown in Figure 10.1. In this form the algorithm does not provide good representations that correspond to very small values of $RM(t)$; however, it is possible to modify the algorithm to dramatically decrease the number of regular multiplications.

The first step is to obtain a suitable approximation of t of the form $2^a F_k^b$, a, b - nonnegative integers. One way to do this is to find a good approximation of $\log t$ of the form

$$\log t \approx a \log 2 + b \log F_k \tag{10.3}$$

This leads us to the theory of linear forms of logarithms, a challenging area in current number theory. The theory, invented by A. Baker [17,18,22], allows us to asymptotically estimate d, the number of binary-Fermat numbers in Equation (10.1). In [15] we proved the following:

Theorem 10.1: Any natural number n can be represented as a sum of $O(\log n/\log\log n)$ numbers of the form $2^a 3^b$.

The proof is based on the following powerful number theoretic result due to Tijdeman [22]:

Theorem 10.2: Let $p_1, p_2, \ldots, p_s$ be a set of s fixed primes. Then there is an absolute effectively computable constant $C > 0$ such that there is always a number of the form $\Pi_{i=1}^{s} p_i^{e_i}, e_i \geq 0$, in the interval $[n-(n/(\log n)^c, n)]$, where n is an arbitrary positive integer.

Input:	t, F_k;
Output:	L = {(a_1,b_1),(a_2,b_2),...,};
Step 0:	i = 0;
Step 1:	i:= i + 1;
Step 2:	Find ; $q_i \le t$; a,b – non-negative integers;
Step 3:	t:= t – q_i;
Step 4:	if t > 0 then go to Step 1 else
Step 5:	Output (L)

Figure 10.1

The greedy algorithm for binary-Fermat numbers.

The proof of Theorem 10.1 can be obtained by applying Theorem 10.2 for the special case $s = 2$, $p_1 = 2$, $p_2 = 3$ and considering the sequence of integers $n = n_0 > n_1 > n_2 > \ldots > n_l > n_{l+1}$ generated by the greedy algorithm.

Since Tijdeman's theorem is valid for the case $s = 2$, $p_1 = 2$, $p_2 = F_k$, then we have the following:

Theorem 10.3: Any natural number, n, can be represented as a sum of $O(\log n/\log\log n)$ binary-Fermat numbers

It is not difficult to see that the result from Theorems 10.1 and 10.3 are the best possible. Both theorems demonstrate that, on average, the DBNS and the binary-Fermat number representations require asymptotically smaller numbers of ones in representing integers as opposed to the binary number system. Therefore, we can expect asymptotic improvements of the performance of algorithms whose complexity depends upon the number of nonzero terms in the input data. From a practical point of view it is also important to have some information about the implicit constant associated with the complexity analysis. Applying current results from number theory, we arrive at a rather pessimistic picture; i.e., all we can say about the constant C, used in Tijdeman's theorem, is that it is smaller than 10^9! Tijdeman's theorem itself produces several counterintuitive conclusions. First, it shows that no matter how many bases are used (as long as their number is fixed and larger than 1), one would have the same asymptotic estimation for the number of summands necessary and sufficient to represent a given integer, n; that is, $O(\log n/\log\log n)$. The proof of Theorem 10.1 [15] can be extended to any given finite set of primes, but the computer experiments with the greedy algorithm with different sets of primes show significantly different performance. This is due to a strange number theoretic phenomenon (pointed out to us by B.M.M. de Weger in a private communication). If we consider the case of primes $p_1 = 17$, $p_2 = 89$, then it turns out that the integers of the form $17^a 89^b$, a, b - nonnegative integers, are distributed in clusters. In such special cases the worst-case behavior of the greedy algorithm might be particularly bad. But the general theory of linear forms of logarithms does not distinguish different choices of primes; that is, a proof that can be applied to the case $p_1 = 2$, $p_2 = 3$ would work equally well in the case $p_1 = 17$, $p_2 = 89$, for example. For this reason one inevitably gets large constants in applying general theorems from the theory of linear forms of logarithms. A special theory devoted to a particular set of primes (say, $p_1 = 2$, $p_2 = 3$) should give us a much more precise picture, but the invention of such a theory appears beyond the reach of the modern transcendental number theory. The computational experiments performed during the work on this chapter and also provided in several papers give us reason to pose the following conjecture:

Conjecture 10.1: Let $p_1, p_2, \ldots, p_s$ be a set of s fixed primes. Then every positive integer, n, can be represented as a sum of $[(2/s)(\log n/\log\log n) + g(n)]$ numbers of the form $\Pi_{i=1}^{s} p_i^{e_i}, e_i \geq 0$, where $[\lim_{n\to\infty}(g(n)\log\log n/\log n) = 0]$.

In the case of DBNS and binary-Fermat number representations, the conjecture posed would allow us to determine that the constant associated with the computational complexity analysis of the greedy algorithm is 1. For more computational experiments, see [15,16,18].

10.3 Finding Suitable Exponents

We try to find a good representation of the exponent, t, according to our needs. As we have mentioned earlier, the use of the greedy algorithm does not, in general, provide good representations. The reason is as follows. If, at some cycle in the algorithm (especially in the first few), the corresponding value of the obtained powers of 2 is low, the power of the odd base would necessarily be high, and thus the total value of $RM(t)$ would be high. Since we can allow ourselves to use very high powers of 2 (they correspond to squarings that are considered free in our computational model), it would seem logical to apply some restrictions on the powers of Fermat numbers used. As a (positive) by-product, these restrictions lead to the simplification and speed up of the greedy algorithm!

Equation (10.1) implies that we have to find a suitable representation of the exponent based on two conflicting conditions:

1. If we allow high values of the powers of the odd base used, then we reduce d, the number of summands, but the maximal power of the odd base will lead to high value of $RM(t)$.
2. If we restrict the maximal power of the odd exponent, then the value of d increases.

We clearly search for the best compromise between these two conflicting conditions. It is quite difficult, if at all possible, to provide a precise compromise; instead, we offer many computational examples to help us in developing heuristic rules, from which one can obtain a suitable exponent representation.

10.3.1 Bases 2 and 3

The easiest case that can serve as a good illustration of the technique proposed above involves the first two prime numbers as bases 2 and 3. So, we look for a representation of the exponent of the form

$$t = \sum_{i,j} d_{i,j} 2^i 3^j, \; d_{i,j} \in \{0,1\} \tag{10.4}$$

where we try to keep j very small while making sure that the number of terms in Equation (10.4) is significantly lower than the Hamming weight of t. Before showing our results from extensive computational experiments with random numbers, we offer an example with a randomly chosen 593-bit integer t (shown in hexadecimal notation).

Example 10.1

t = 14B2726E AC5C24D FFA1D093C714341C 82507C33DFE111B3
2E7408BDD349CD0 0A61CE0EF10C41A5 CDB32DE23CA5BCD
444A967B1665DEA8D 05CBAE2B7BD39 7D08E2F58A3058F7A1
A43F2183

The representation of t in the DBNS with no restriction on the ternary exponent (in this case the maximal possible exponent of 3 might be as high as $\lfloor 592/\log_2 3 \rfloor = 373$) is shown in Table 10.1. The representation of the same

TABLE 10.1

Index Representation of t as a Sum of Numbers of the Form 2^a3^b with No Restrictions on the Powers of 3, That is, $t = 2^{315}3^{175} + 2^{501}3^{52} + 2^{426}3^{93} + \ldots + 2^{8}3^{2} + 2^{3}3^{3} + 2^{0}3^{1}$

315, 175	501, 52	426, 93	279, 188	102, 286	112, 274	63, 299	392, 86
451, 52	342, 105	437, 39	154, 211	237, 153	199, 170	372, 55	260, 120
380, 38	211, 139	164, 162	327, 53	18, 241	80, 197	76, 193	44, 205
74, 181	121, 146	264, 50	319, 9	58, 168	24, 184	283, 16	107, 120
162, 79	193, 53	117, 95	169, 55	73, 107	196, 25	44, 114	85, 83
40, 105	42, 98	18, 108	57, 79	44, 83	123, 29	11, 94	49, 63
33, 69	17, 75	129, 0	58, 39	61, 33	75, 20	76, 15	44, 31
6, 51	67, 8	11, 39	57, 5	6, 33	49, 2	11, 22	16, 15
9, 16	3, 16	11, 8	5, 8	8, 2	3, 3	0, 2	

TABLE 10.2

Index Representation of t as a Sum of Numbers of the Form 2^a3^b with Largest Ternary Exponent 17, That is, $t = 2^{567}3^{16} + 2^{568}3^{11} + 2^{563}3^{10} + \ldots + 2^{13}3^0 + 2^73^1 + 2^03^1$

567, 16	568, 11	563, 10	549, 16	570, 0	536, 17	532, 16	534, 12
547, 0	528, 0	517, 12	514, 11	501, 16	508, 8	489, 17	505, 3
489, 9	473, 15	470, 14	465, 12	476, 1	469, 1	448, 9	434, 15
436, 11	438, 7	416, 17	425, 7	419, 7	397, 17	394, 13	386, 14
392, 6	370, 14,	371, 8	368, 6	348, 15	353, 9	336, 15	329, 16
334, 10	336, 6	317, 14	318, 10	305, 15	313, 7	297, 14	310, 2
298, 5	278, 13	273, 11	249, 16	257, 7	248, 8	227, 13	216, 17
210, 16	226, 3	218, 2	203, 7	198, 7	181, 15	189, 6	180, 8
172, 10	172, 7	152, 16	148, 15	154, 7	142, 10	152, 1	131, 9
126, 9	113, 14	99, 13	104, 7	81, 17	86, 17	84, 9	86, 3
65, 12	68, 7	60, 9	44, 16	48, 10	39, 13	41, 7	32, 10
31, 6	34, 1	26, 2	14, 5	14, 2	13, 0	7, 1	0, 1

number in the form Equation (10.4) with maximal ternary exponent 17 is shown in Table 10.2. This particular choice of the maximal ternary exponent, that is, 17, will be explained later.

The number of regular multiplications, $RM(t)$, in the first case (no restrictions on the ternary exponent) is 299 (the largest ternary exponent participating in the DBNS representation) plus 71 (the number of summands of the form 2^a3^b) minus 1, that is, 369. This is in sharp contrast to the second case, where we have 17 (the largest allowed ternary exponent) plus 96 (the number of summands of the form 2^a3^b) minus 1, that is, 112.

Our next two goals are:

1. To show how one can use, as efficiently as possible, the representation scheme proposed;
2. To demonstrate why the use of small ternary exponents leads to so drastic a reduction of the number of regular multiplications.

10.4 Algorithm for Exponentiation with a Low Number of Regular Multiplications

In order to make the proposed representation as suitable as possible for efficient exponentiation, the pairs (binary-exponent, ternary-exponent) have to be reordered in increasing order of the ternary exponent. For the number used in Example 10.1, this particular reorder is shown in Table 10.3.

TABLE 10.3

Reordered Representation of the Exponents Shown in Table 10.2

Ternary Exponent	The Corresponding Exponents of 2, in Increasing Order, Participating in the Representation of t	Number of Exponents of 2
0	13, 547, 570	3
1	0, 7, 34, 152, 469, 476	6
2	14, 26, 218, 310	4
3	86, 226, 505	3
4	—	0
5	14, 298	2
6	31, 189, 336, 368, 392	5
7	41, 68, 104, 154, 172, 198, 203, 257, 313, 419, 425, 438	12
8	180, 248, 371, 508, 528	5
9	60, 84, 126, 131, 353, 448, 489	7
10	32, 48, 142, 172, 318, 334, 563	7
11	273, 436, 514, 568	4
12	65, 465, 517, 534	4
13	39, 99, 227, 278, 394	5
14	113, 297, 317, 370, 386, 470	6
15	148, 181, 305, 336, 348, 434, 473	7
16	44, 152, 210, 249, 329, 501, 532, 549, 567	9
17	76, 81, 216, 397, 416, 489, 536	7

This reordered representation is described by the following equation:

$$t = \sum_{i=0}^{m} 3^i \left(\sum_{j=1}^{c(i)} 2^{b_i^{(j)}} \right) \tag{10.5}$$

where m is the maximal ternary exponent used; $c(i)$ is the number of binary exponents that correspond to the ith ternary exponent, $0 \le i \le m$; and $b_i^{(j)}$ is the jth binary exponent that corresponds to the ith ternary exponent, $1 \le j \le c(i)$. Note that in some cases $c(i)$ might be zero.

With the above notations in hand, we now propose the algorithm shown in Figure 10.2.

In order to clarify the details, some comments on selected steps in the algorithm are provided here.

Step 1 sets the two registers (R_0) and (R_1). For example, if the computations are performed over $GF(2^n)$, then they are two variables of type $GF(2^n)$.

Input: $A, t, m, c(i), b_i^{(j)}$;
Output: $B = A^t$;
Step 0: $i = 0$;
Step 1: $(R_0) = (R_1) = A; j = 1; k = 0$;
Step 2: **while** $(j \le c(i))$ **begin**
Step 3: $(R1) = (R1)^{2b_i^{(j)}}$;
Step 4: **if** $(k = 0)$ **begin** $(B) = (R_1); k = 1$ **end**
else $(B) = (B) * (R_1)$
Step 5: $j = j + 1$ **end**;
Step 6: $i = i + 1$;
if $(i \le m)$ **begin**
$(R_1) = (R_0) * (R_0); (R_0) = (R_1) * (R_0); (R_1) = (R_0)$; **goto** *Step* 2;
end
else
Step 7: **Output** (B)

FIGURE 10.2
Algorithm for exponentiation.

Step 3 performs a cyclic shift of the current value of the register (R_1).

Step 4 updates the value of (B), the register that will contain at the end of the algorithm the final result. We save one multiplication by observing that the first multiplication to update (B) can be replaced by an assignment of the multiplier to (B), since (B) is equal to 1 at this point.

Step 6 explicitly cubes the value of register (R_0), that is, $(R_0) = A^{3^i}$, $0 \le i \le m$, and assigns it to (R_1). The final result is contained in (B).

Note that step 4 requires $\Sigma_{i=0}^{m} c(i) - 1 = d - 1$ regular multiplications, and step 6 requires m regular multiplications. By applying the above algorithm to the 593-bit number used in Example 10.1, one obtains 112 regular multiplications (96 in Step 4, and 16 in Step 6). This is substantially better than the known methods (129 in Stinson's algorithm, which is regarded as the fastest one [24]).

The proposed algorithm is particularly attractive if the computational operations are performed over a large finite field with even characteristic. The use of a normal base representation of the elements of the field allows us to implement the squarings (see Steps 3 and 6 of the algorithm) as cyclic shifts, which can be implemented with a low VLSI cost function. The ECC encryption technology, invented by Certicom Corp., uses exactly these algorithmic considerations to achieve significant computational speed up of the ECC encryption/decryption procedures.

10.5 Complexity Analysis Using Exponential Diophantine Equations

The average complexity of the proposed method depends on the way the number of terms of the form 2^a3^b in Equation (10.4) decreases (in the average case) as a function of the largest ternary exponent allowed. For the sake of clarity, consider that the exponent, t, is a 593-bit integer. As we mentioned earlier, the maximal exponent of 3 that can appear in a double-base number representation of t is 373. This choice corresponds to an unrestricted size of the ternary exponent and it is clearly unsuitable for our goals. The minimal exponent of 3 is, of course, zero, which corresponds to a purely binary representation. In this case the average number of ones is 297 and the expected number of regular multiplications is 296. In the case of an unrestricted ternary exponent, Theorem 10.1 indicates that the expected number of terms of the form 2^a3^b in Equation (10.4) is $\lfloor 593/\log_2 593 \rfloor = 64$. Therefore, the upper bound of the maximal number of regular multiplications is 436. Clearly, neither binary nor a purely double-base technique can outperform Stinson's method [20], which uses only 129, on average, regular multiplications (a more complete computational analysis is available in [24]). But Example 10.1, Table 10.2 demonstrates that by allowing only very small ternary exponents one can sharply decrease the number of terms of the form 2^a3^b in Equation (10.4). To understand why this is the case, we will use some information from the theory of exponential Diophantine equations.

First, let us consider some very small values of m, that is, the largest ternary exponent allowed to be used.

Case 1

First of all, let us consider the case $m = 0$. This is the purely binary representation of the exponent, t. In this case half of the nonzero digits (bits) are expected to be ones. Therefore, the average number of regular multiplications is $\lfloor \log_2 t/2 \rfloor$.

Case 2

This is the first nontrivial case, $m = 1$; ie., the largest ternary exponent allowed of 1. If one applies the greedy algorithm, then the combination of numbers 2^k and 2^{k+1} cannot occur because they will be replaced by $3 \cdot 2^k$. With no restriction on the ternary exponent, every solution of the Diophantine equation,

$$x + y = z;\ GCD(x, y, z) = 1;\ x, y, z \in \{2^a3^b;\ a, b - \text{nonnegative integers}\} \quad (10.6)$$

would produce an impossible combination of numbers (x, y). Equation (10.6) has exactly three solutions [16]: (1, 2, 3), (1, 3, 4), and (1, 8, 9). If the ternary exponent is restricted to be no larger than 1, then only the first solution plays a role. For this particular case, that is, the largest ternary exponent of 1, we can prove the following:

Theorem 10.4: Let t be a positive integer that is represented via the greedy algorithm in the following form:

$$t=\sum_{i=1}^{d}3^{a(i)}2^{b(i)};\ a^{(i)}\in\{0,1\};\ b^{(i)}-\text{ nonnegative integers} \tag{10.7}$$

Then the average value of d, the number of summands in (10.7), is $\lfloor \log_2 t/3 \rfloor$.

Proof: A single *isolated* bit 1 occurs in the binary representation of t with probability 1/8 (corresponds to a succession of bits 010). *Exactly two* isolated consecutive bits 11 occur in the binary representation of t with probability 1/16 (corresponds to a succession of bits 0110). The greedy algorithm will reduce the two ones with one number of the form 3.2^k. If we have three consecutive 1 bits in the binary representation of t, then the greedy algorithm will replace the two most significant ones with one number of the form 3.2^k. Generally, if we have l consecutive ones in the binary representation of t, then the greedy algorithm will reduce them to $l/2$ terms of the form $3{\cdot}2^k$, if l is even. If l is odd, then the greedy algorithm will reduce the number of nonzero binary digits to $(l–1/2)$ terms of the form $3{\cdot}2^k$ and one power of 2 will remain unchanged. Thus, if t is represented in the form of Equation (10.7), then the average value of d, the number of summands of the form $3^{a^{(i)}}2^{b^{(i)}}$; $a^{(i)} \in \{0, 1\}$, $b^{(i)}$ – non-negative integers, is given by the following sum:

$$d=\left\lfloor\left(\frac{1}{8}+\frac{1}{16}+\frac{2}{32}+\frac{2}{64}+\frac{3}{128}+\frac{3}{256}+\ldots\right)\log_2 t\right\rfloor$$

$$=\left\lfloor\left(\sum_{k=2}^{\infty}\frac{3(k-1)}{2^{2k}}\right)\log_2 t\right\rfloor=\left\lfloor\frac{\log_2 t}{3}\right\rfloor$$

which completes the proof

The above analysis shows that one can achieve significant savings in terms of nonzero digits in the DBNS even if the largest ternary exponent allowed is only 1. More to the point, one gets exactly the same reduction (33%) of the nonzero digits that is achieved in the binary signed digit (SD) number representation but with a 33% reduction in the size of the digit set (the DBNS digit set is {0, 1} versus {–1, 0, 1} for the SD number representation). As we shall see from the computational experiments with random numbers, the estimate obtained in Theorem 10.1 is in very good agreement with the numerical results.

By increasing the size of the ternary exponent allowed, a much larger class of exponential Diophantine equations and their solutions start to play a role. For instance, if the ternary exponent is bounded by 2, then the solution (1, 8, 9) provides another impossible combination of numbers (x, y). Generally speaking, every solution of the exponential Diophantine equation,

$$x_1 + x_2 + \ldots + x_{k-1} = x_k\,;\; GCD(x_i) = 1;\; i = 1,2,\ldots,k \tag{10.8}$$

in numbers of the form $2^{a_i}3^{b_i}$, $i = 1, 2, \ldots, k$, a_i, b_i – non-negative integers, generates some sort of reduction rule, which starts to play a role in reducing the number of nonzero digits so long as the largest ternary exponent allowed is greater than or equal to the largest b_i, $i = 1, 2, \ldots, k$. The solutions [16,22] of the equation

$$x_1 + x_2 + x_3 = x_4;\; x_1, x_2, x_3, x_4 \in \{2^a 3^b\};\; GCD(x_1, x_2, x_3, x_4) = 1 \tag{10.9}$$

can be found in Table 2.3, and repeated for convenience in Table 10.4.

Another equation that plays a role in analyzing the complexity of the greedy algorithm with reduced ternary exponent is

$$x_1 + x_2 + x_3 = x_4 + x_5;\; x_1, x_2, x_3, x_4, x_5 \in \{2^a 3^b\};\; GCD(x_1, x_2, x_3, x_4, x_5) = 1 \tag{10.10}$$

To the best of our knowledge, Equation (10.10) has not yet been completely solved, although some investigations have been reported [18]. Our computational experiments suggest that this equation possesses about 500 different solutions. However, in this particular case some solutions can be excluded since they do not lead to the reduction of three numbers of the form 2^a3^b to two if one uses the greedy algorithm. The smallest example showing this state of affairs is 41; in this case the greedy algorithm returns $41 = 36 + 4 + 1$, whereas $41 = 32 + 9$ is the minimal representation of 41. The identity $36 + 4 + 1 = 32 + 9$ shows a solution, such that $\max(x_1, x_2, x_3) > \max(x_4, x_5)$. The solutions having this property do not produce needed reductions. However, only about 10% of the solutions found by our computational experiments possess this property.

The most general class of exponential Diophantine equations that can be considered in this case consists of the following equations:

TABLE 10.4

Solutions of the Diophantine Equation (10.9)

1, 2, 3, 6	1, 2, 6, 9	1, 2, 9, 12	1, 2, 24, 27	1, 3, 4, 8	1, 3, 8, 12
1, 3, 12, 16	1, 3, 32, 36	1, 4, 27, 32	1, 6, 9, 16	1, 8, 9, 18	1, 8, 18, 27
1, 8, 27, 36	1, 8, 72, 81	1, 9, 54, 64	1, 12, 243, 256	1, 16, 64, 81	1, 27, 36, 64
1, 32, 48, 81	1, 256, 512, 729	2, 3, 4, 9	2, 3, 27, 32	2, 9, 16, 27	3, 4, 9, 16
3, 8, 16, 27	8, 9, 64, 81				

$$\sum_{i=1}^{k} x_i = \sum_{j=1}^{l} y_j;\ x_i, y_j \in \{2^a 3^b\};\ GCD(x_1, x_2, \ldots, x_k, y_1, y_2, \ldots, y_l) = 1;\ k > l \tag{10.11}$$

One of the most profound results in modern transcendental number theory asserts that the number of solutions of Equation (10.11) is finite. The proven upper bound on the number of solutions is a double exponential function of the number of variables, $k + l$; the proven lower bound is a single exponential function of $k + l$ [18]. This is one of the reasons why (1) it is very difficult to find all the solutions, and (2) it is probably impossible to thoroughly analyze their influence on the performance of the greedy algorithm.

10.6 Experiments with Random Numbers

Now we offer some information based on very extensive simulations with 593-bit integers. Our goals are:

1. To show the way the number of terms of the form $2^a 3^b$ in Equation (10.4) decreases as the largest ternary exponent allowed increases;
2. To find the optimal value of the largest ternary exponent for the problem in hand.

Table 10.5 shows the results from this computational experiment. For every particular value of the maximal ternary exponent we performed 1,000,000 experiments with randomly chosen 593-bit integers. The calculations were performed with the help of NTL—a library for performing number theory (which is available from the University of Wisconsin). Table 10.5 shows the values of the average number of terms of the form $2^a 3^b$ in Equation (10.4) for randomly chosen 593-bit integers and the corresponding number of the number of regular multiplications, $RM(t)$, as a function of the largest ternary exponent allowed, t. Table 10.5 shows the corresponding values only for very small t, ($1 \le t \le 4$), around the optimal value, which turns out to be equal to 17 ($15 \le t \le 19$) and some larger values ($29 \le t \le 32$). The computational experiments were performed for every value of t between 1 and 60.

As can be seen from Table 10.5, by carefully choosing the value of the largest ternary exponent, one can dramatically reduce the number of regular multiplications in performing exponentiation. It also numerically confirms the result encapsulated in Theorem 10.1, that is, if the largest ternary

TABLE 10.5

Results from Computational Experiments

Largest Ternary Exponent, t	Average Number of Terms of the Form 2^a3^b in Equation (11.4)	Corresponding Value of $RM(t)$
1	198.36	198.36
2	173.36	17.36
3	154.01	153.01
4	138.6	137.6
…	…	…
15	102.48	116.48
16	101.41	116.41
17	100.2	116.2
18	99.42	116.42
19	98.55	116.55
…	…	…
29	90.26	118.26
30	89.64	118.64
31	88.95	118.95
32	88.5	119.5

exponent is 1, then the expected number of regular multiplications is equal to one-third of the binary length of the exponent.

10.6.1 Other Bases

We have performed essentially the same experiments with two other sets of binary-Fermat numbers: 2^a5^b, and 2^a17^b. There is no particular difference from an implementation viewpoint, so we just report here the final conclusions about the most important figures of merit—optimal maximal exponents of the odd bases and the corresponding average value of the number of regular multiplications. Again, all the experiments performed have used 593-bit exponents.

For the set of bases 2 and 5, the optimal value of the exponent of the quintary base turns out to be 24, and the corresponding average number of regular multiplications is 117.87.

For the set of bases 2 and 17, the optimal value of the exponent of 17 turns out to be 26, and the corresponding average number of regular multiplications is 121.63.

As it is seen, in general these two new sets of binary-Fermat numbers do not outperform the DBNS, but since the difference is marginal (in favor of the DBNS), it is worth checking every particular exponent to choose the best representation.

10.7 A Comparison Analysis

There are two major figures of merit to be considered: time (in this case, equivalent to the number of regular multiplications) and area—defined as the number of registers used to calculate modular exponentiation. The second criterion is especially important in smart card applications, where there are severe restrictions on the size of the memory allowed. There are several algorithms aimed at minimizing the number of regular multiplications to calculate modular exponentiation, among which the algorithm proposed by Stinson [20] is regarded as the fastest. Walter's algorithm, based on division chains [21], also has very good performance, and it is particularly attractive due to its low storage requirements. Finally, the binary method [12] is not only the best known and widely used, but it is very efficient in terms of area. Table 10.6 summarizes the comparisons among the different methods and figures of merit; the size of the exponent is 593 bits.

10.8 Final Comments

Some final comments on the proposed technique might be helpful for the reader in understanding the circumstances under which the demonstrated algorithm can be successfully applied.

1. The algorithm is best suited if the calculations are performed over a large finite field with even characteristic. Canadian-based Certicom Corp. is producing cryptoprocessors that operate over such finite fields.
2. The algorithm does require very few registers, which makes it very suitable for hardware implementation, particularly in the case of severe memory restrictions.
3. One of the best techniques used to improve the performance of modular exponentiation is the use of signed digit number representations [9,10,23,24]. In the case of elliptic curves, it is justified by the fact that division is replaced by subtraction. We can generalize Equation

TABLE 10.6

Comparison between Different Exponentiation Algorithms

Algorithm	Binary [12]	Stinson [20]	Walter [21]	Bases 2, 3	Bases 2, 5	Bases 2, 17
RM	297	129	>140	116	118	122
No. registers	2	31	3	3	3	3

(10.4) to the digit set {–1, 0, 1}, which would lead to a further reduction of the number of summands, *d*, and would diminish the number of regular multiplications.

4. Some of the digital signature algorithms use multiexponentiations at a certain stage of the computational process. The most common example is the verification procedure of the digital signature in the Digital Signature Standard (DSS). While the calculation of multi-exponentiation, with as few as possible modular multiplications, has been a subject of considerable research [29,36,37], we are unaware of any particular results devoted to minimizing the number of *regular* multiplications in this case. This problem certainly seems worth researching.

10.9 Summary

In this concluding chapter, we have considered how the use of multiple-base number representations—properly applied—can lead to significant improvements in implementing a very basic cryptographic computational operation, namely, exponentiation over finite field with even characteristic. The bottom line, again, is that one can achieve very sparse double-base representations by making use of a very small subset of powers of the odd base. Indeed, the main theoretical advantage of the multiple-base representation (in comparison to their single-base positional counterparts) is that the number of nonzero elements used is asymptotically smaller. In this chapter we have seen that this property remains valid even if we use a very small portion of the powers of the odd base. Since elliptic curve cryptographic systems constitute a significant part of mobile security systems (e.g., the encryption routines used by Research In Motion (RIM) in their Blackberry™ devices almost exclusively apply ECC technology), the algorithms proposed in this chapter may find their commercialization pathway. Extensions to fields with small odd characteristics deserve attention, particularly because, in this case, one has to minimize the largest binary exponent used. The reader may find these and other open theoretical and practical problems sufficiently appealing to explore.

References

1. R. Rivest, A. Shamir, L. Adleman, A Method for Obtaining Digital Signatures and Public Key Cryptosystems, *Communications of the ACM*, 21, 120–126, 1978.

2. W. Diffie, M. Hellman, New Directions in Cryptography, *IEEE Transactions on Information Theory*, 22, 644–654, 1976.
3. A.J. Menezes, *Elliptic Curve Public Key Cryptosystems*, Kluwer Academic, Dordrecht/Norwell, MA, 1993.
4. V. McLellan, Password Security-Crypto in Your VAX, *Digital Review*, October 1986, p. 86.
5. T. ElGamal, A Public Key Cryptosystem and a Signature Scheme Based on Discrete Logarithms, *IEEE Transactions on Information Theory*, 31, 469–472, 1985.
6. E.F. Brickell, K.S. McCurley, Interactive Identification and Digital Signatures, *AT&T Technical Journal*, 74–86, 1991.
7. C.P. Schnorr, Efficient Identification and Signatures for Smart Cards, in *Advances in Cryptology*, Crypto '89, vol. 435, pp. 239–252, Lecture Notes in Computer Science, Springer Verlag, Berlin, 1989.
8. G.B. Agnew, R.C. Mullin, S.A. Vanstone, Fast Exponentiation in GF(2^n), in *Advances in Cryptology*, Eurocrypt '88, vol. 330, pp. 251–255, Springer Verlag, Berlin, 1988.
9. F. Morain, J. Olivos, Speeding Up Computations on an Elliptic Curve Using Addition-Subtraction Chains, *Information Theory and Applications*, 24, 531–543, 1990.
10. K. Koyama, Y. Tsuruoka, A Signed Binary Window Method for Fast Computing Over Elliptic Curves, *IEICE Transactions on Fundamentals* (Special Issue on Cryptography and Information Security), E76-A, 55–62, 1993.
11. J. Chao, K. Tanada, S. Tsujii, Design of Elliptic Curve with Controllable Lower Boundary of Extension Degree for Reduction Attack, in *Advances in Cryptology*, Crypto '94, 1994, pp. 50–55.
12. D.E. Knuth, Seminumerical Algorithms, in *The Art of Computer Programming*, vol. 2, Addison Wesley, Reading, MA, 1981.
13. J. Olivos, On Vector Addition Chains, *Journal of Algorithms*, 2, 13–21, 1981.
14. V.S. Dimitrov, T.V. Cooklev, Two Algorithms for Modular Exponentiation Using Non-Standard Arithmetic, *IEICE Transactions on Fundamentals* (Special Issue on Cryptography and Information Security), E78-A, 82–87, 1995.
15. V.S. Dimitrov, G.A. Jullien, W. C. Miller, An Algorithm for Modular Exponentiation, *Information Processing Letters*, 66, 155–159, 1998.
16. V.S. Dimitrov, G.A. Jullien, W.C. Miller, Theory and Applications of the Double-Base Number System, *IEEE Transactions on Computers*, 48, 1098–1106, 1999.
17. A. Baker, *Transcendental Number Theory*, Cambridge University Press, Cambridge, 1975.
18. B.M.M. de Weger, Algorithms for Diophantine Equations, *CWI Tracts—Amsterdam*, 65, 1989.
19. R. Tijdeman, On the Maximal Distance between Integers Composed of Small Primes, *Compositio Mathematicae*, 28, 159–162, 1974.
20. D.R. Stinson, Some Observations on Parallel Algorithms for Fast Exponentiation over GF(2^n), *SIAM Journal on Computing*, 19, 711–717, 1990.
21. C. Walter, Exponentiation Using Division Chains, *IEEE Transactions on Computers*, 47, 757–765, 1998.
22. R.J. Stroeker, R. Tijdeman, Diophantine Equations, *CWI Tracts—Amsterdam*, 155, 321–369, 1987.

23. E.F. Brickell, D.M. Gordon, K.S. McCurley, D.B. Wilson, Fast Exponentiation with Precomputation, in *Advances in Cryptology*, Eurocrypt '92, vol. 658, pp. 200–207, Springer Verlag, Berlin, 1992.
24. D.M. Gordon, A Survey on Fast Exponentiation Methods, *Journal of Algorithms*, 27, 129–146, 1998.
25. N. Koblitz, Elliptic Curve Cryptosystems, *Mathematics of Computation*, 48, 203–209, 1987.
26. P. Erdös, Remarks on Number Theory, III, on Addition Chains, *Acta Arithmetica*, 6, 77–81, 1960.
27. Y. Tsuruoka, A Fast Algorithm for Addition Sequences, in *Proceedings of Korea-Japan Joint Workshop on Information Security and Cryptology '93*, 1993, pp. 70–73.
28. A. Miyagi, On Secure and Fast Elliptic Curve Cryptosystems over F_p, *IEICE Transactions on Fundamentals*, E77-A, 630–635, 1994.
29. A.C. Yao, On Evaluation of Powers, *SIAM Journal on Computing*, 5, 70–73, 1976.
30. J. von zur Gathen, Efficient and Optimal Exponentiation in Finite Fields, *Computational Complexity*, 1, 360–394, 1991.
31. Y. Tsuruoka, K. Koyama, Fast Exponentiation Algorithms Based on Batch Processing and Precomputations, *IEICE Transactions on Fundamentals* (Special Issue on Cryptography and Information Security), E80-A, 34–39, 1997.
32. Ch. Paar, P. Fishermann, P. Soria-Rodriguez, Fast Arithmetic for Public-Key Algorithms in Galois Fields with Composite Exponents, *IEEE Transactions on Computers*, 48, 1025–1034, 1999.
33. G. Agnew, R. Mullin, S. Vanstone, An Implementation of Elliptic Curve Cryptosystems over GF(2^{155}), *IEEE Journal of Selected Areas in Communications*, 11, 804–813, 1993.
34. E. DeWin, A. Bosselaers, S. Vanderberghe, P.D. Gersem, J. Vandewalle, A Fast Software Implementation for Arithmetic Operations in GF(2^n), in *Proceedings of Asia*, Crypt '96, 1996, pp. 65–76.
35. J. Guajardo, C. Paar, Efficient Algorithms for Elliptic Curve Cryptosystems, in *Advances in Cryptography*, Crypto '97, 1997, pp. 342–356.
36. V.S. Dimitrov, G.A. Jullien, W.C. Miller, Complexity and Fast Algorithms for Multiexponentiations, *IEEE Transactions on Computers*, 49, 92–98, 2000.
37. S.-M. Yen, C.-S. Laih, A. Lenstra, Multiexponentiation, *IEE Proceedings—Computers and Digital Techniques*, 136, 325–326, 1994.

Index

C

D